# R. McNEILL ALEXANDER
# HUMAN BONES
## A SCIENTIFIC AND PICTORIAL INVESTIGATION

PHOTOGRAPHY *by* AARON DISKIN

CONSULTANT: HENRY GALIANO

A PETER N. NEVRAUMONT BOOK
PI PRESS · NEW YORK

PI PRESS

An Imprint of Pearson Education, Inc.
1185 Avenue of the Americas, New York, New York 10011

Pi Press offers discounts for bulk purchases. For information contact U.S. Corporate
and Government sales, 1-800-382-3419, or corpsales@pearsontechgroup.com.
For sales outside the U.S.A., please contact International Sales, 1-317-581-3793,
or international@pearsontechgroup.com.

Printed in P.R. China.

First Printing

Library of Congress Cataloging-in-Publication Data
A CIP catalog record for this book can be obtained from the Library of Congress

Pi Press books are listed at www.pipress.net

ISBN 0-13-147940-7

Pearson Education Ltd.
Pearson Education Australia Pty., Limited
Pearson Education Singapore, Pte. Ltd.
Pearson Education North Asia Ltd.
Pearson Education Canada, Ltd.
Pearson Educación de Mexico, S.A. de C.V.
Pearson Education — Japan
Pearson Education Malaysia, Pte. Ltd.

Produced by Nevraumont Publishing Company, New York, New York 10006
Jacket and Book Design: Matthew Schwartz

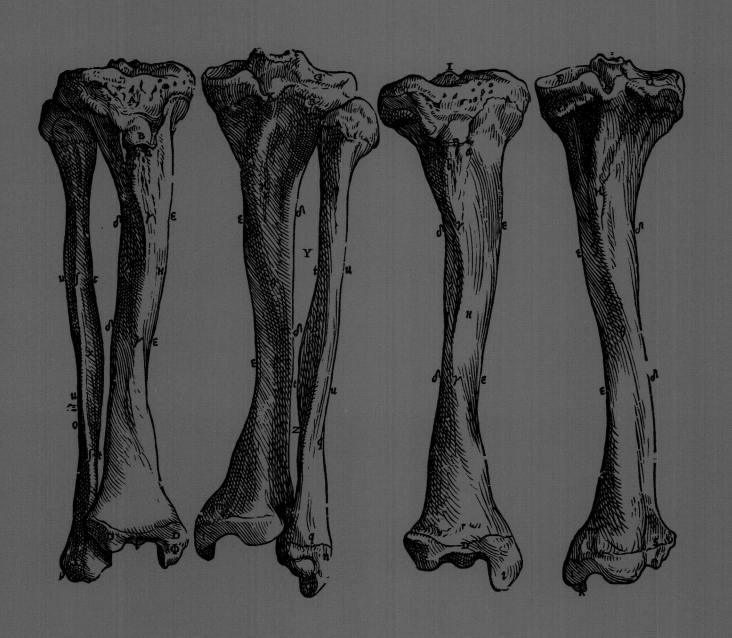

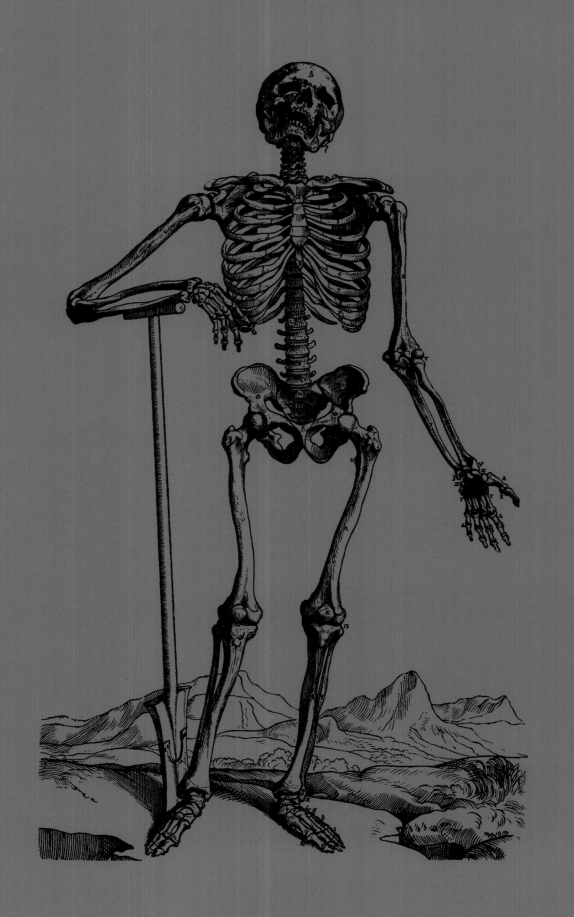

# CONTENTS

PLATE 1
*An adult human skeleton seen from the side.*
*(Courtesy Maxilla & Mandible.)*

PLATE 2
*The skeleton from in front.*
*(Courtesy Maxilla & Mandible.)*

# INTRODUCTION

The bones of a human skeleton can tell us a great deal about how our bodies are constructed, and about how they work. [Plates 1 and 2] Bones, however, come from dead bodies and many of us would be very reluctant to look at a corpse. We tend to be even more unwilling to examine a body cut open to show the muscles and other organs inside, despite the understanding that this might provide of our own living bodies.

Dry bones, though, are different. With the soft, messy parts of the body removed, they seem to have left life and death so far behind them that we can look at them and even handle them without fear or nausea. We are free to learn from them, and to admire their intriguing and beautiful shapes. The elegance of bones is fully revealed in this book, never before as colorfully and dramatically as in the photographs by Aaron Diskin that are showcased here.

Throughout the course of our exploration, particular bones will be referred to by name. Figure 1 shows the names of many of the human body's 213 bones. The names of the individual bones that are sutured together to form the skull, and the small carpal and tarsal bones of the wrist and ankle, have been left out of the figure. Including them would have made it intolerably cluttered, and knowing their names is not necessary for an understanding of this book.

We'll begin by looking at bones as living, growing organs. Next is a tour of the human skeleton—the skull, the arms and legs, and the torso. We'll move from the skeleton to a discussion of diseased and damaged bones, and from there to a look at why the bones of different people are different. In the last chapter, we'll explore why bones are the way they are in form and function—that is, the evolution of human bones. It is a story of how we came to be standing here.

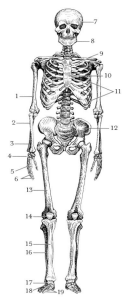

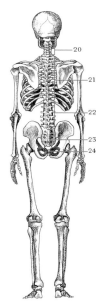

*Figure 1 Front and rear view of human skeleton: (1) humerus, (2) radius, (3) ulna, (4) carpals, (5) metacarpals, (6) phalanges, (7) skull, (8) lower jaw, (9) clavicle, (10) sternum, (11) ribs, (12) pelvic bone, (13) femur, (14) patella, (15) tibia, (16)fibula, (17) tarsals, (18) metatarsals, (19) phalanges, (20) cervical vertebrae, (21) thoracic vertebrae, (22) lumbar vertebrae, (23) sacrum, (24) coccyx*

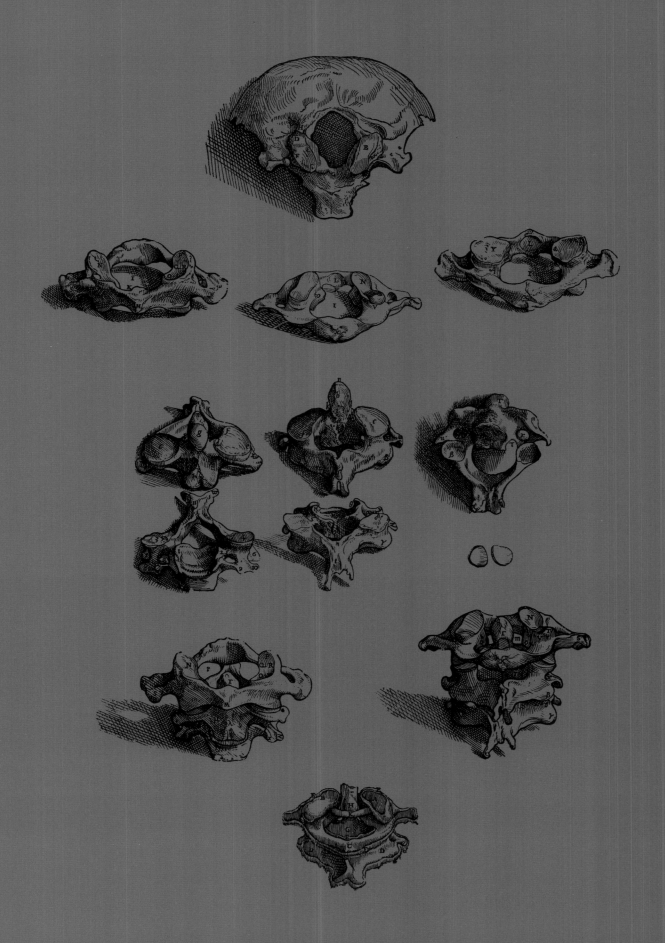

# 1

# LIVING, GROWING BONES

Nothing seems as lifeless as a bone in a museum case, but the bones in our bodies are as truly alive as our muscles and guts. The life is in the cells, which are scattered throughout the bone, each in its tiny individual cavity. Like other living cells, they need energy to keep them alive. Some of the energy from the food that we eat is used by our bones, and some of the oxygen that we breathe is used in the bones to release the energy from the food.

## ENERGY

Many years ago, in the 1950s, two physiologists (A. W. Martin and F. A. Fuhrman) set out to discover how much energy was used by each of the tissues of a resting animal. They took thin slices of tissue from many different parts of the bodies of anesthetized dogs (which were eventually killed without being allowed to recover from the anesthetic). They put these samples into flasks containing a little dog blood serum and kept them at a normal dog body temperature. They measured the rate at which each sample used oxygen.

Martin and Fuhrman found that different tissues used oxygen at very different rates, showing that some of them used energy much faster than others. For example, skin used energy at a rate of 1 milliwatt per gram, muscle used 3 mW/g, brain 8 mW/g and kidney a massive 14 mW/g. Remember that these are resting rates; the muscle in particular would have used energy very much faster, when contracting actively in an intact animal. Bone used only 0.2 mW/g. I estimate, however, from microscope

sections of bone that the cells occupy only about 1-2% of its volume. Allowing for the fact that cells are less dense than the rest of the bone, they would be 1% or less of the weight of the bone. This makes bone very different from tissues such as muscle, which consist mainly of cells, with only a little fluid in the intercellular spaces. Calculating the rate of energy use per gram of cells leads to the conclusion that bone cells are among the most active in the body. This chapter will show what keeps them busy.

## BLOOD

Many of the cells lie deep within the bones. There is no question of oxygen diffusing in from the bone surface fast enough to keep the bones alive. Blood vessels are needed throughout the bone, to keep the cells supplied with oxygen and foodstuffs. The slices of bone in the experiment were able to survive without a functioning blood supply, only because they were cut very thin.

Arteries and veins enter bones through small holes such as the ones in the neck of the femur (the thigh bone). [Plate 3] Some of their branches go right through into the marrow, but others divide into capillaries that run, parallel to each other, throughout the bone itself. Microscope sections show how the bone is organized around the network of capillaries. [Figure 2] Several different arrangements are found in different kinds of bone, but most human bone is built of units called osteons. Each osteon is a cylinder of 0.1 to 0.4 mm diameter, with a capillary running along its axis in a canal (called a Haversian canal). The bone is laid down in concentric layers about 3 micrometers (0.003 mm) thick, around the canal. The cells lie in spaces between the layers. Spidery extensions of the cells reach out through fine pores in the bone to connect the cells to each other and to the Haversian canal. This network of connections conveys oxygen and foodstuffs from the blood capillaries to the cells. You may wonder whether there is any advantage in having energy-consuming cells in our bones. Are they not an unnecessary drain on our resources? Mollusk shell has no cells, and works reasonably well as a skeleton. If we had no cells in our bones, we would save a little energy.

## BONE REPAIR

One big advantage of cells is that they can repair damaged bones. Most of us break bones very seldom, if at all, but we suffer a great deal of damage of which we are generally not aware. This is due to the phenomenon of bone fatigue, which has nothing to do with feeling tired. "Fatigue" is used here in a technical sense quite different from its meaning in everyday language.

Metal fatigue causes a great deal of worry to engineers, and has been responsible for some spectacular disasters. It has been blamed for steel ships breaking in half, and

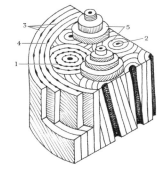

Figure 2 *Diagram of a wedge cut from a bone with parts of the lamellae stripped away to show how fiber directions alternate in successive lamellae. (1) bone cells, (2) Haversian canal, (3) lamellae, (4) surface bone, (5) osteons.*

Plate 3 *(opposite) The upper end of a femur (thigh bone). The holes in the neck of the bone, just below its head, are passages for the blood vessels that are needed to keep the bone alive. (Courtesy Maxilla & Mandible.)*

for a recent helicopter crash due to a rotor blade breaking. Large stresses may be needed to break a piece of metal in a one-off test, but much lower stresses repeated thousands or millions of times may eventually lead to failure. Each repetition causes a minute amount of damage, which accumulates. The same phenomenon has been demonstrated in tests on pieces of bone, using the machines that engineers use for similar tests on metals and plastics. There are unexplained differences between the results of experiments performed by different scientists in slightly different ways. A typical result is that a stress of half of what would be needed to break a bone in a one-off test will cause fatigue failure after a million or so applications.

Stresses of that order probably act in the tibia in running, as I will show in Chapter 5. (The tibia is the principal bone of the lower leg; see Figure 1). An athlete who runs 100 km each week takes nearly two million running strides each year, stressing each tibia once in every stride. That sounds dangerous. Unlike the bones in the engineering tests, however, athlete's bones have living cells in them. These generally work fast enough to repair the damage before there is any danger of the bone breaking. Mollusks cannot repair damage except at the edges of their shells, but the cells in our bones can repair damage wherever it occurs.

Just occasionally the repair mechanism proves inadequate. Athletes occasionally get shin splints, which is fatigue damage in the tibia. The shin feels sore, and tiny cracks can sometimes be detected in X-ray pictures of the tibia. The cure is to stop running for a while and rest, avoiding any activity that would impose large stresses on the bone. If the patient does that, the natural repair process will generally restore the bone to its normal strength. If he or she continues running, however, the cracks will grow and the bone may eventually break. Human athletes generally do not let the damage progress that far, but racehorses sometimes break their legs in circumstances that clearly indicate fatigue damage. Typically, a leg bone breaks and the horse collapses without warning, while the horse is galloping on the flat. There is no preliminary stumble, and the bone has clearly broken before the horse falls.

The processes that keep our bones in good repair are constantly active, removing old bone and replacing it with new. Specialized cells eat away the bone around a blood capillary, and then other cells build new bone in the empty space. The cells around one capillary may eat away some of the osteon around the next capillary, so that a section through the bone shows a mosaic of osteons of different shapes, some complete cylinders and some very incomplete.

The processes that repair bones are not limited to restoring them to normal strength. They also strengthen bones or allow them to weaken, in response to changes in the demands that we put on them. New recruits in armies have often developed fatigue fractures in the metatarsal bones in their feet, presumably because they have suddenly

*Plate 4 (opposite) One person's forearm bones. Corresponding bones from the left and right sides of the body may not be exact mirror images of each other. (Courtesy Maxilla & Mandible.)*

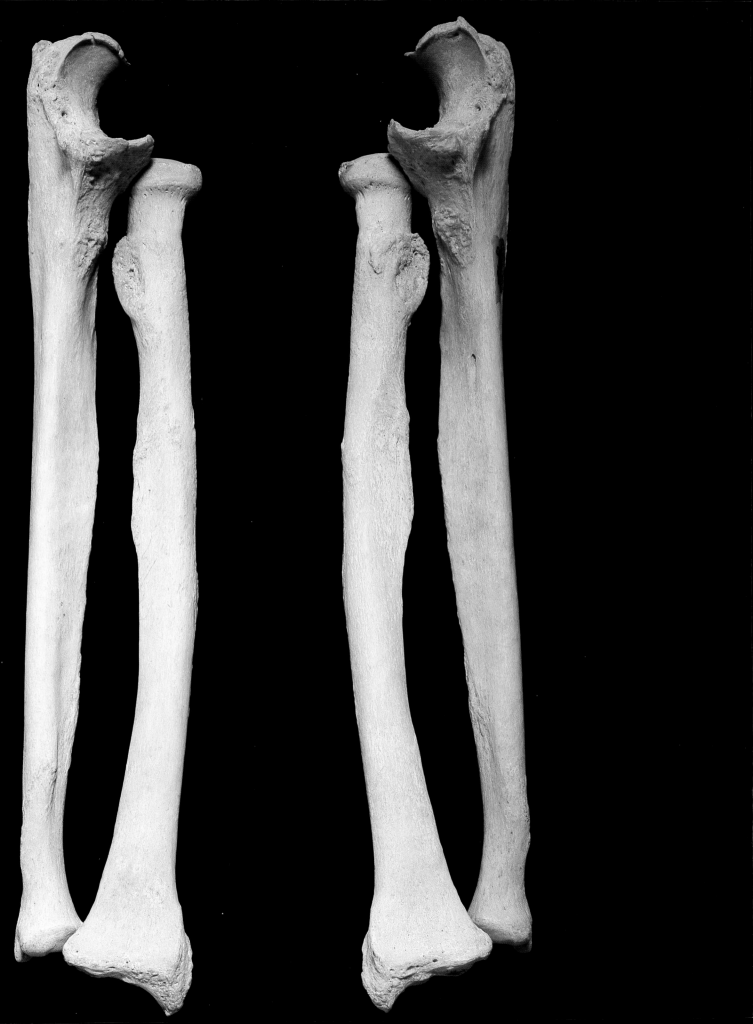

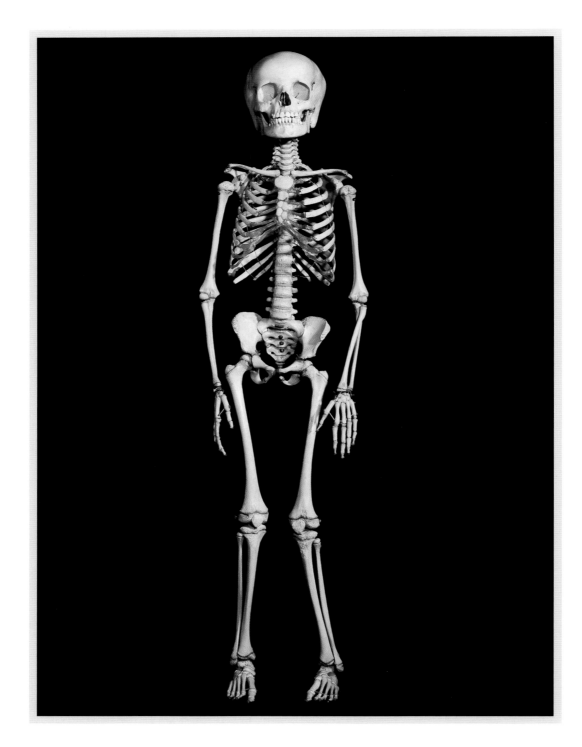

PLATE 5
*The skeleton of an eight-year-old Eskimo girl.*
*(Courtesy G. J. Sawyer.)*

started spending a lot of time marching. The metatarsals will eventually strengthen enough to withstand the frequently repeated stresses of army life, but if the change of lifestyle is too sudden they may not respond fast enough, and the bones may crack.

## BONE STRENGTH AND WEAKNESS

Whether right-handed or left-handed, most people use one arm more than the other. The preferred arm may develop stronger bones, but often there is no very obvious difference in thickness. [Plate 4] Tennis players, however, exert large forces much more often with one arm, than with the other. X-ray images of the arms of professional tennis players have shown that the bones are thicker in the arm that holds the racquet. Changes also occur in the opposite direction. Bones become weaker in patients who are confined to bed, and in astronauts who spend substantial times in reduced gravity.

The processes that we have been discussing strengthen bone or allow it to weaken, in response to use or disuse. More dramatic changes occur in growing children and young animals. American boys double in stature between the ages of 2 and 18 years. Girls at the peak of their growth spurt, at age 12, grow at a rate of about 8 cm/year. Boys in their slightly later growth spurt at age 14 grow at 10 cm/year. Growth rates are even higher, 15-20 cm/year, in the first few months after birth. As stature increases, bones of course get longer. One study showed the length of the tibia increasing from averages of 19 cm (girls) and 18 cm (boys) at age 3, to adult lengths of 36 cm (girls) and almost 40 cm (boys). Growth ceased at age 16 (girls) and 18 (boys).

## BONE GROWTH

The growth of a child is not simply a matter of getting bigger. Shape and proportions change as well. Children's skulls are larger, relative to the rest of the skeleton, than the skulls of adults. (Compare Plate 5 with Plate 2). They are also rounder than in adults, with a relatively smaller face. [Plate 6]

Dwarfs are people whose bones have not grown normally. [Plate 7] In the commonest form of dwarfism, known as achondroplasia, the arm and leg bones remain unusually short relative to the head and torso. Achondroplasia results from a mutation in a single gene, which affects about one birth in 30,000.

Early studies of bone growth were made by feeding a harmless dye to pigs. Madder is the dye used to produce the beautiful pink and red colors in Persian carpets. It binds to calcium, and if it is present in the blood while new bone is being formed it dyes that new bone pink. Pigs were given madder in their food for a while and killed some time later. When the leg bones were sawn in half, a pink layer was revealed in the cross section. If the pig had been killed immediately after the period of feeding

*Plate 6 (following pages) Skulls of a 4 to 5 year old child (left) and an adult (right). The adult's braincase is little larger than that of the child, but the bones of the face are distinctly larger. (Courtesy Division of Anthropology, American Museum of Natural History.)*

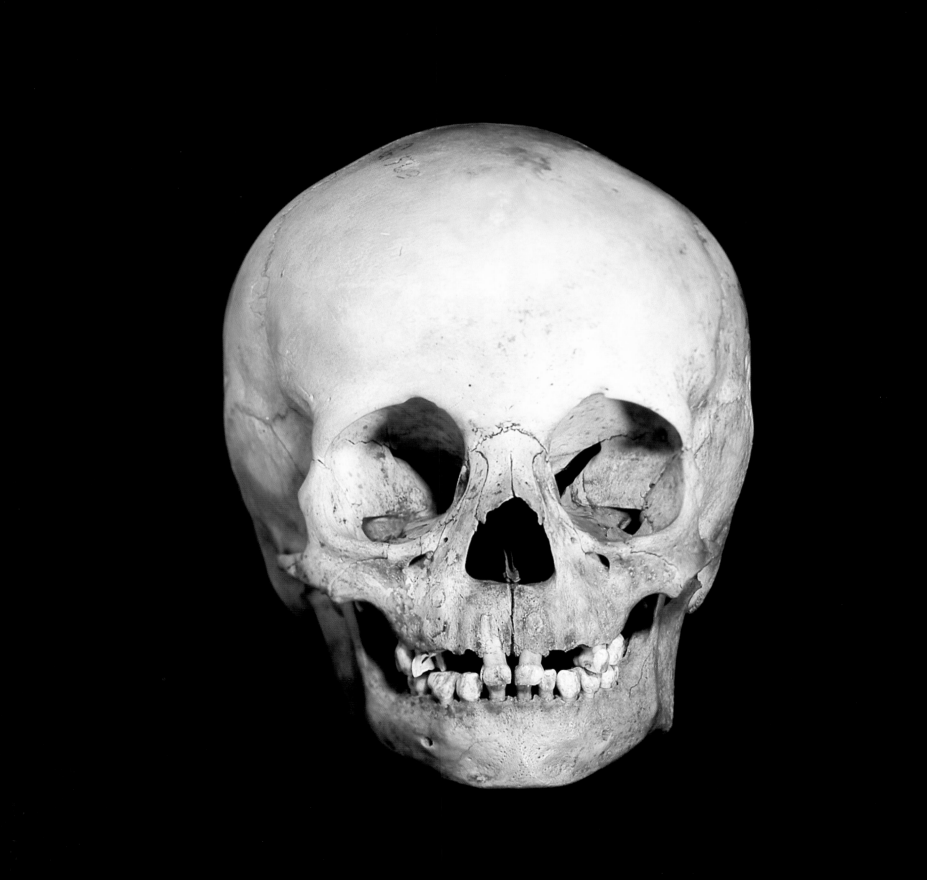

with madder, the pink color was at the outer surface of the bone. If it had been kept alive for some time after madder feeding ceased, the pink was deeper in the bone. Eventually it reached the marrow cavity and disappeared. The experiment showed that the bone got thicker by adding new bone on the outside, and removing the old bone on the inside, next to the marrow cavity. The diameter of the marrow cavity increases proportionately, as the bone gets thicker. That pattern of growth means that none of the material in a young piglet's leg bones is retained in the adult pig.

If skulls were seamless like eggshells they would have to grow in the same way, simply by adding new bone on the outside and removing bone on the inside. All of the bone would have to be replaced every time the radius of the skull increased by an amount equal to its thickness. But skulls are not seamless; they consist of 19 separate bones. That makes a much more economical pattern of growth possible. New bone is added at the edges of the individual bones as well as to their thickness. Bone has to be removed in some places while it is being added in others because the bones have to become less sharply curved as they grow, but much less turnover of bone is needed than if the skull had no sutures.

You might imagine that leg bones would grow in length by adding new material at their ends, but it is not as simple as that. Instead, new bone is added a little short of the ends. A child's limb bones are not continuous bone from end to end, but have three parts. The main shaft of the bone, the diaphysis, is capped at each end by small pieces of bone called epiphyses. [Figure 3 and Plate 8] At joints, the epiphysis of one bone contacts the epiphysis of the next, through the layers of articular cartilage which aid lubrication. Each epiphysis is connected to the diaphysis by another thin layer of cartilage, the epiphysial plate.

Growth in length happens at the epiphysial plates, where new bone is added to the ends of the diaphyses. Some bone has to be added all round the growing epiphyses to keep their shape as they get larger, but there is much less growth close under the articular cartilage of the joint surfaces than there would be if all the growth occurred there. It has sometimes been suggested that this is advantageous, that the epiphyses enable the bone to grow with minimal disturbance of the joints. Lizards, however, seem to manage perfectly well without epiphyses.

Epiphysial plates are regions of potential weakness. One possibility is that a sideways force might shear an epiphysis off the end of the bone. This might be a serious danger if epiphysial plates were flat, but they are generally very bumpy, with humps on the surface of the diaphysis that fit into hollows in the epiphysis and vice versa. [Plate 9] With surfaces that interlock like that, the epiphysis is less likely to be sheared off. When children reach adult size and stop growing, their epiphysial plates disappear, so that the epiphyses and diaphysis become fused together as a single piece of bone. [see Plate 8]

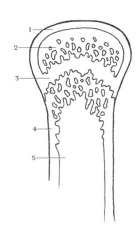

*Figure 3 Head of bone. (1) articular cartilage, (2) epiphysis, (3) epiphysial plate, (4) diaphysis, (5) marrow cavity.*

*Plate 7 The leg bones of a dwarf compared to those of a full-sized adult. (Courtesy Maxilla & Mandible.)*

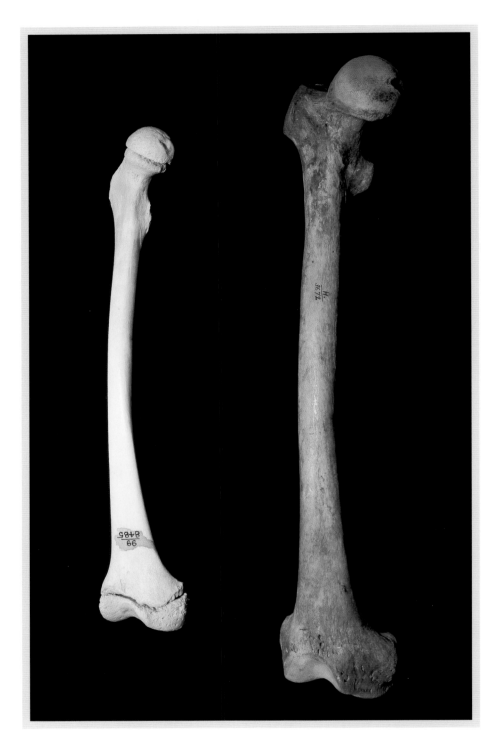

PLATE 8
*Femurs of a teenager (left) and an adult (right). The epiphyses of
the younger bone are still not fused to the shaft.
(Courtesy Division of Anthropology, American Museum
of Natural History.)*

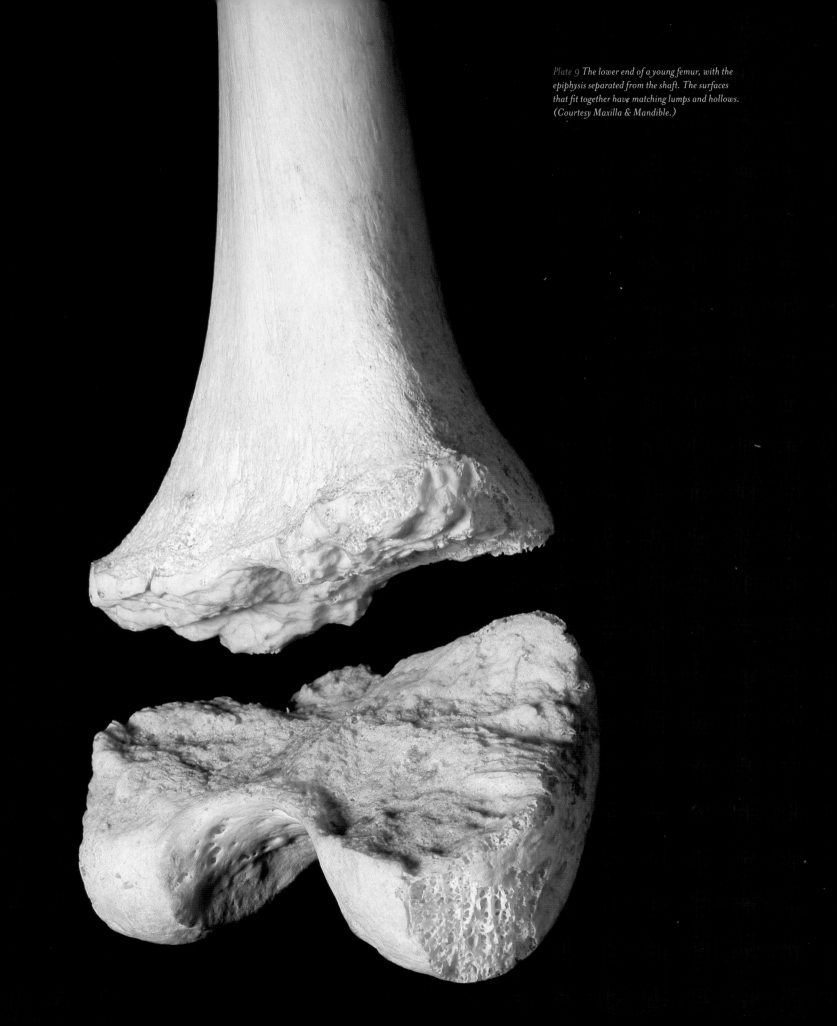

Plate 9 *The lower end of a young femur, with the epiphysis separated from the shaft. The surfaces that fit together have matching lumps and hollows. (Courtesy Maxilla & Mandible.)*

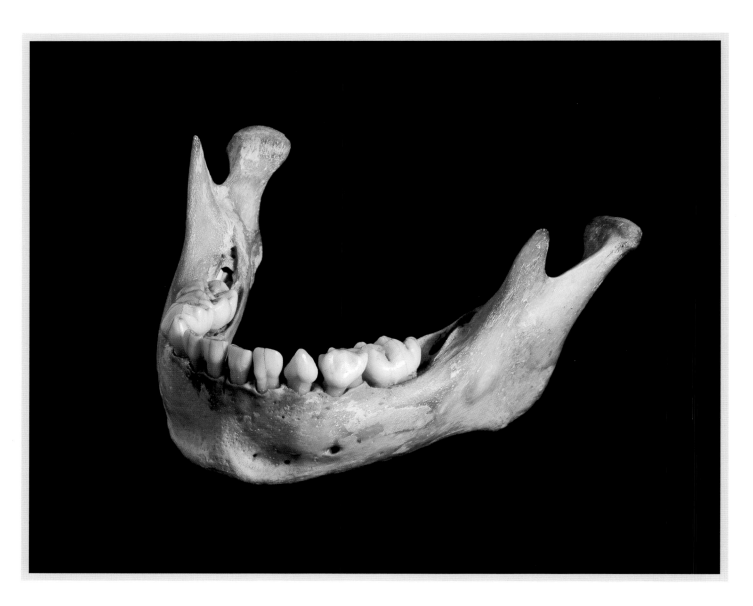

PLATE 10
*A child's lower jaw. The milk teeth are all in place,
and the first adult molar can be seen erupting on the far
side of the mouth. (Courtesy Division of Anthropology,
American Museum of Natural History.)*

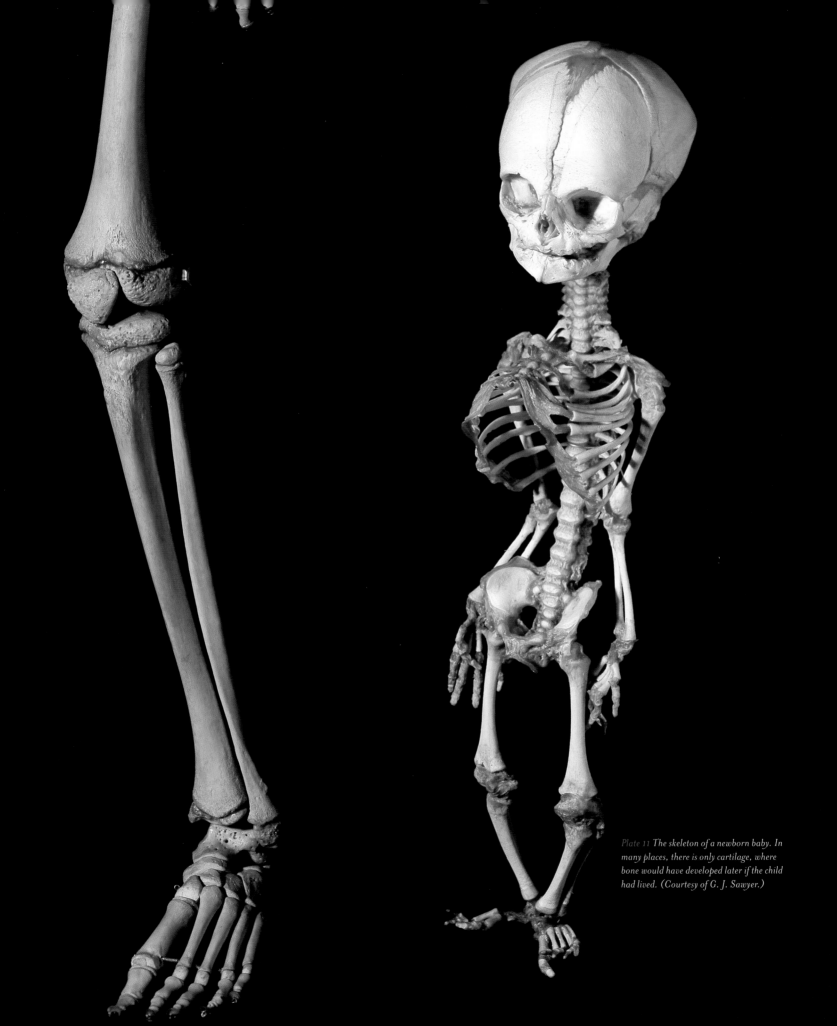

*Plate 11 The skeleton of a newborn baby. In many places, there is only cartilage, where bone would have developed later if the child had lived. (Courtesy of G. J. Sawyer.)*

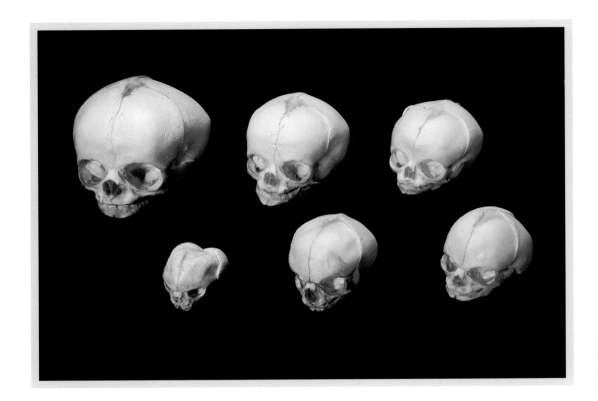

*Plate 12 A sequence of skulls, showing stages of development before birth. (Courtesy Division of Anthropology, American Museum of Natural History.)*

*Plate 13 A baby's skull. There is a gap between the bones of the skull roof. (Courtesy Maxilla & Mandible.)*

## TEETH

Unlike their bones, young children's teeth do not grow to adult size. Just as we discard our clothes and shoes as we grow, replacing them with larger sizes, we discard and replace our first teeth. Unlike clothes, we replace teeth only once. Adult teeth are larger than the milk teeth of young children, but not sufficiently larger to fill the mouth of a full-grown adult if there were no additional teeth. As well as the 20 teeth that are direct replacements for milk teeth, adults have 12 additional teeth at the back of the mouth: three molars on each side of each jaw.

The milk teeth erupt more or less in order from the front backwards, in the first 2.5 years of life. The canines commonly appear out of order, after the teeth immediately behind them. The permanent teeth start appearing at age 6 or 7 years, starting with the first incisors and the first molars. [Plate 10] The last teeth to erupt are the third molars or wisdom teeth. They do not appear until the young adult has stopped growing, usually between 17 and 21 years. We do not have a full set of 32 teeth in place until we have a full-sized jaw to hold them.

Apart from the difference in size, the skeletons of new-born babies are very different from those of adults. [Plate 11] They have cartilage (gristle) in many places where adults have bone. There is only cartilage in the wrist where adults have many small carpal bones. There are tiny pieces of bone that will become epiphyses at the lower end of the femur and the upper end of the tibia, but all the other long bones end in lumps of cartilage. Many more bony epiphyses appear during the first year of life, but the set is not complete until an age of about five years.

## BABY BONES

Babies' skulls grow very rapidly, both before and after birth. [Plate 12] At birth, the head is one quarter of the total length of the body, whereas in adults it is only one eighth of stature (compare Plate 11 with Plate 2). It has to be disproportionately big to hold the brain, which makes up 10% of the weight of the baby compared to only 2% in the adult. With an average mass of 350 grams, the baby's brain is already one quarter of adult size. Though it starts big, the brain grows fast and reaches 90% of adult size by the age of five. The eyes of new-born babies are in about the same proportion to brain size as in adults, but the rest of the face and the jaws are relatively much smaller.

New-born babies have thin skull bones. The bones do not meet at the top of the head, but are joined by a strong leathery membrane. [Plate 13] This structure allows fast growth, but has its dangers. The skull roof is flexible, and easily distorted by anything that presses against it. Even lying always in the same position may alter the shape of the skull. Pressure on the sides of the skull may make it abnormally long

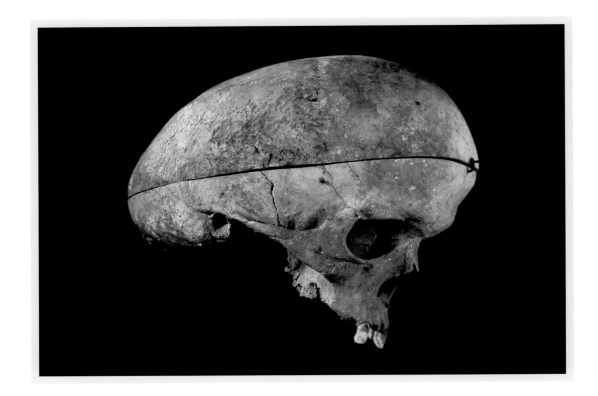

*Plate 14 A scaphocephalic skull.*
*(Courtesy Henry Galiano.)*

*Plate 15 A plagiocephalic skull.*
*(Courtesy Henry Galiano.)*

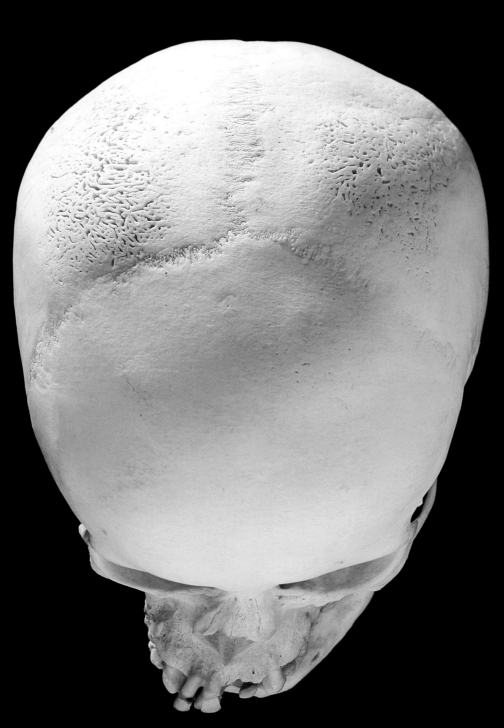

*Plate 16* **A microcephalic skull.**
*(Courtesy Henry Galiano.)*

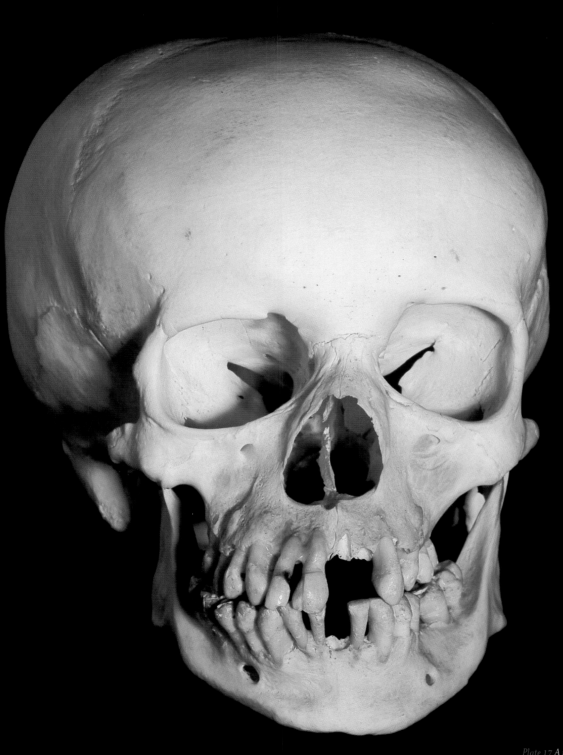

*Plate 17* A microcephalic skull.
(Courtesy Henry Galiano.)

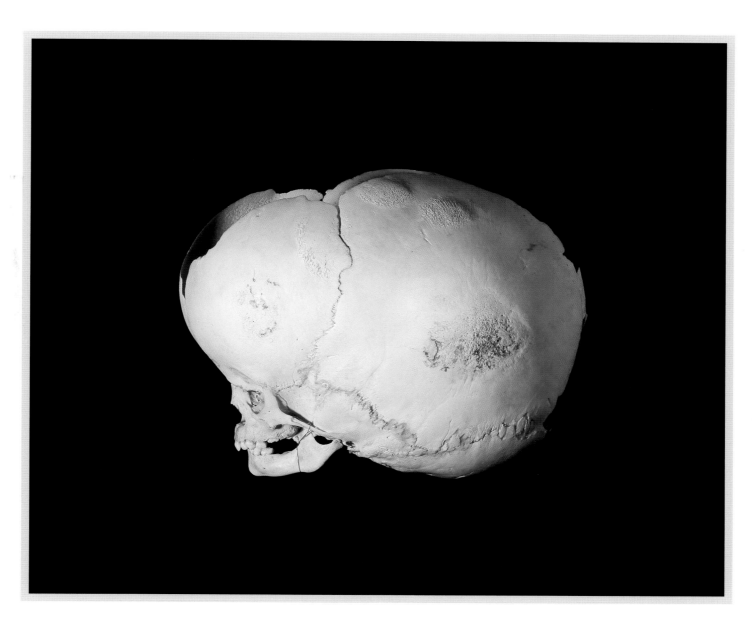

PLATE 18
*An extreme case of hydrocephaly, in a child's skull.*
*(Courtesy Henry Galiano.)*

and narrow, the condition known as scaphocephaly. [Plate 14] Pressure on one side of the back of the skull may make it asymmetrical, the condition of plagiocephaly. [Plate 15] Happily, the flexibility that allows the skull to distort also makes correction possible. A suitably shaped head band or helmet will usually mould the growing skull to a normal shape within a few months. Scaphocephaly and plagiocephaly can, however, result from some of the bones meeting and suturing together too soon, making it difficult for the growing skull to expand in some directions. In these cases, surgery may be needed. In normal growth, the gaps between the bones are not fully closed until the child is about a year old.

The growth of babies' skulls seems to be driven by the growth of the brain inside. In the rare disorder known as microcephaly, the brain fails to grow as it should. Though the face grows to normal size, the rest of the skull does not. [Plates 16 and 17] Common results include mental retardation and early death. In contrast, in hydrocephaly far too much fluid accumulates in the cavities of the brain. The brain swells, and the skull grows with it. [Plates 18 and 19] This condition can be treated by surgery designed to let the excess fluid drain away. Both microcephaly and hydrocephaly are often associated with abnormalities of the chromosomes.

This chapter has shown that the bones of living people, like their other tissues, are very much alive. The living cells in bones, and the food and oxygen brought to bones in the blood, enable children's bones to grow and adults' bones to maintain themselves. Occasionally something goes seriously wrong, but usually the life processes work well. The next few chapters will look more closely at individual bones.

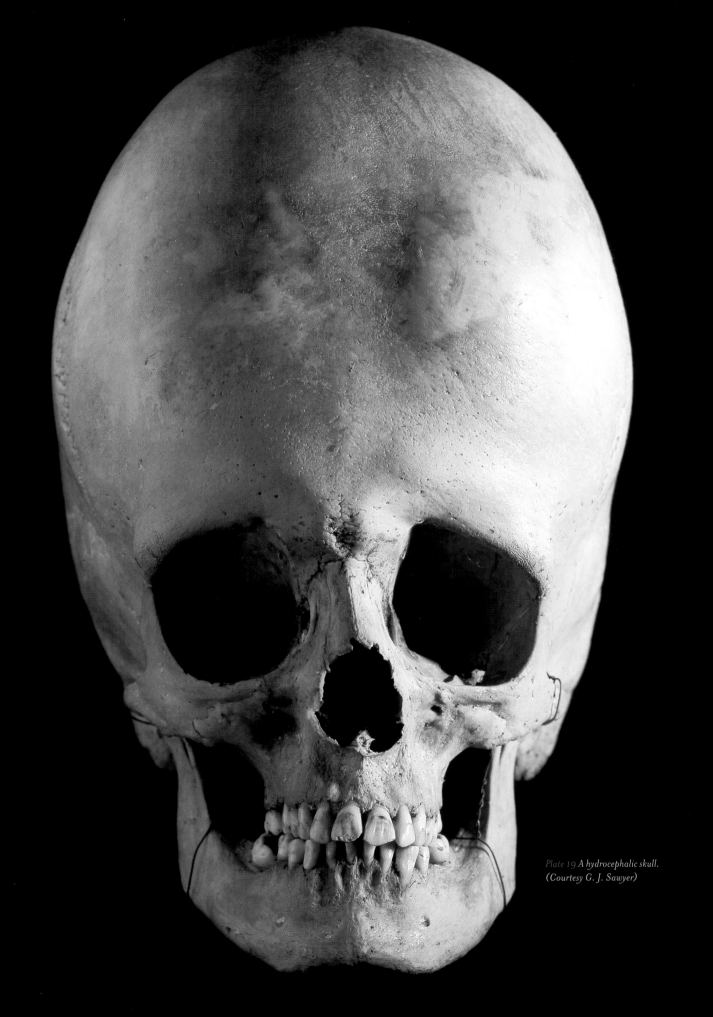

*Plate 19 A hydrocephalic skull.
(Courtesy G. J. Sawyer)*

# 2

# THE SKULL

We will start our survey of the skeleton with the skull [Plate 20], which is formed from many separate bones. Most of these bones remain separate throughout life, but the two bones in a baby's forehead [see Plate 13] almost always unite to form a single bone in adults. Where the bones meet, they are connected by interlocking sutures, like exceedingly complicated pieces from a jigsaw puzzle. [Plate 21] Collagen fibers hold the sutures together, attaching each bone very firmly to its neighbors. Collagen is the protein that also forms the deeper layers of the skin, the membranes that wrap many of the organs of the body and the tendons that connect muscles to bones.

## FRACTALS

These wiggly sutures between the bones of the skull are long, and it is impossible to say how long. You could use a measuring tape to get the straight-line length between one end of a suture and the other, but that measure would ignore the extra length due to the wiggles. You could lay a thread along the suture, matching the wiggles as best you could, then pull the thread straight and measure it. That would give a greater length, and you would get a still greater one, about three times the straight-line length, if you could measure all the tiny wiggles that are visible in the photograph. Even that would not give a definitive length, because a microscope would reveal even smaller wiggles. Lines like this are studied by mathematicians, in the branch of their subject called fractal geometry. Coastlines are another example.

## SKULL BONES

The brain occupies most of the space inside the skull, but it is not a tight fit. Membranes between it and the bone provide some cushioning and help to prevent it from being shaken around too much by blows to the head. The plates of bone that enclose the brain are sandwiches of two kinds of bone: two sheets of compact (non-porous) bone with a layer of spongy bone between them. Spongy bone is a network of slender bars of bone, with marrow filling the spaces between them. Sandwich structure makes the skull bones much less easily dented than if the same amount of bone were made into a single compact sheet. Similar sandwich construction is used by engineers to make structures such as the walls of aircraft, which need to be strong and stiff but nevertheless light in weight. Corrugated cardboard is another example of sandwich construction. The sandwich principle is effective because the surface layers of a plate of bone or metal are best placed to resist bending forces. Removing material from the middle layers saves weight, with little effect on bending strength. Enough filling must be left in the sandwich to prevent it from collapsing inwards, but there is scope for considerable saving of materials.

The bones of the skull roof have grooves on their inner surfaces, like river systems seen on a map. [Plate 22] These are the paths of some of the blood vessels that serve the brain and its membranes. A good blood supply is needed to bring oxygen and foodstuffs to the brain because it uses a remarkable amount of energy. The brain makes up only 2% of the mass of the body, but it accounts for 20% of the energy consumption of the resting body.

## HUMAN BRAIN

The human brain is remarkably large, in comparison with the brains of animals. Its mass averages 1.4 kg in men, and 1.3 kg in women: the difference is simply due to men having bigger bodies than women. In contrast, a sheep with the body mass of an adult woman would have a brain of only about 0.1 kg. Even a woman-sized chimpanzee would have a brain of only about 0.5 kg. Elephants and baleen whales have larger brains than us, but that is not surprising in animals that are so much larger than we are. The serious challenge to our claim to be the brainiest species comes from dolphins that are about the same weight as ourselves but have heavier brains. Their advanced social behavior and their use of echolocation may require a big brain, although they have nothing like human intelligence. [Figure 4]

The cerebral hemispheres are by far the largest part of the human brain and fill the upper part of the braincase. Their surfaces are deeply wrinkled, but there is no sign of the wrinkles on the smooth inner surface of the roof of the skull. [see Plate 22] The cerebral hemispheres seem to be principally responsible for our most remarkable talents, our capacity for abstract thought and communication.

*Plate 20 (opposite) A human skull. (Courtesy Maxilla & Mandible.)*

36

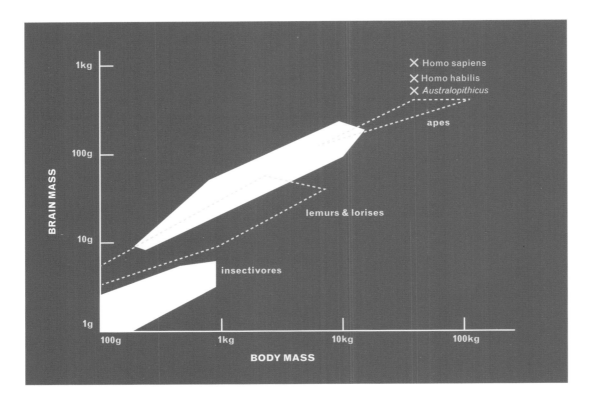

*Figure 5 The relationship of brain mass to body mass in mammals.*

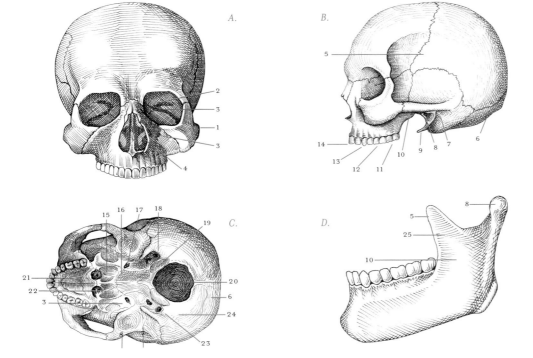

*Figure 6 Four views of the skull and jaw. Bone openings for nerves and attachment areas of muscles are also labeled.*

*A. (1) zygomatic arch, (2) optical nerve, (3) part of the trigeminal nerve, (3) part of the trigeminal nerve, (4) turbinals*

*B. (5) temporalis muscle, (6) neck muscle, (7) ear, (8) jaw articulation, (9) hyoid bone attaches here, (10) masseter muscle, (11) molars, (12) premolars, (13) canine, (14) incisors*

*C. (15) medial pterygoid muscle, (16) lateral pterygoid muscle, (17) facial nerve, (18) vagus nerve, (19) articulation with atlas, (20) foramen magnum (21) palate, (22) nasal cavity, (3) part of trigeminal nerve, (8) jaw articulation, (7) ear, (23) internal carotid artery, (24) digastric muscle, (6) neck muscles*

*D. (5) temporalis muscle, (25) coronoid process, (8) jaw articulation, (10) masseter muscle*

Our understanding of these parts of the brain is still very incomplete. It seems fundamentally impossible that we will ever understand it fully; there must be limits to the extent to which a brain can understand itself. Some of the most remarkable recent advances have been made by magnetic resonance imaging (MRI) and positron emission tomography (PET). These techniques produce images of the brains of living subjects, showing which parts of the brain are active by detecting their increased metabolism. They have shown, for example, that there is a region at the back of the cerebral hemispheres that is concerned with vision. [Figure 5] In PET studies, this part of the brain lights up when the subject is asked to read a word. A different region on the side of the hemisphere lights up when the subject hears a spoken word, and yet another region, further up the hemisphere, when the subject speaks. This area, which apparently controls the movements of speech, is followed further up the side of the hemisphere by areas that control movements of other parts of the body. The sizes of the areas reflect the complexity of control that is required. The organs of speech, and the hands, are quite small parts of the body, but have big areas on the brain to control them.

Figure 5 with its labels on different parts of the cerebral hemispheres, may remind us of phrenology. [Plate 23] The pseudoscience of phrenology, however, associated parts of the cerebral hemispheres with qualities of character such as cautiousness, firmness and benevolence. What modern science has shown is that different parts of the hemispheres are concerned with different senses, or with movements of different parts of the body.

The cerebral hemispheres make up three-quarters of the brain. The remaining quarter is in the lower parts of the braincase. The complexity of these lower parts, with their many distinct subdivisions, is reflected in the shape of the floor of the braincase. [Plate 24] The cerebellum [see Figure 5] is responsible for the coordination of movement. The medulla controls vital functions such as breathing and the heartbeat. It is connected directly to the spinal cord through the foramen magnum, the hole in the base of the skull. [Plate 25] Knobs of bone on the back of the skull, on either side of the hole, rest on the first neck vertebra. The neck muscles attach to the bone surrounding the hole.

## EYE SOCKETS

The eye sockets are shaped to fit the spherical eyeballs. The eyeballs need to be spherical, to be able to tilt up and down, and to turn from side to side, to look in any direction. In addition to these movements, the spherical shape of the eyeball allows it to roll in its socket while continuing to look straight forward. You cannot do that voluntarily, but there is a reflex that makes the eye roll in this way when the head is tilted, so as to keep the image in the same place on the retina. Six muscles connecting the eye socket to different points on the eyeball drive these movements.

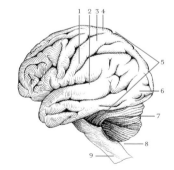

*Figure 5 The brain with labels indicating the roles of some parts of the cerebral hemispheres. (1) speech, (2) hearing, (3) arm, (4) torso, (5) cerebral hemisphere, (6) vision, (7) cerebellum, (8) medulla, (9) spinal cord*

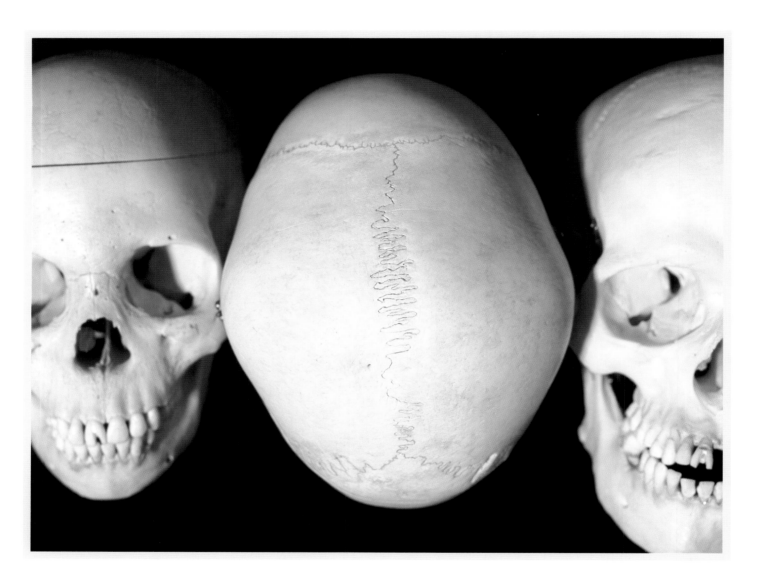

PLATE 21
*The bones of the skull roof interlock like jigsaw pieces.*
*(Courtesy Maxilla & Mandible.)*

The size of the eye sockets shows that the eyes are considerably larger than they seem to be when you see them half hidden by eyelids, in a living head. They need to be big to see fine detail, because the bigger an eye, the more sense cells can be fitted onto the retina. The light-sensitive cells at the centre of the retina, where they are most closely packed, are about 2.5 micrometers (2.5 thousandths of a millimeter) apart. They cannot be packed more tightly because the cells apparently cannot be made smaller; the cells are the same size in a mouse retina as in a human one. The diameter of the human eyeball is about 24 millimeters, so the retina is about 24 mm behind the lens, about ten thousand times the distance between adjacent sense cells. That implies that the very smallest detail we can hope to be able to distinguish at a distance x is one ten thousandth part of x. We can hope to be just able to see one millimeter details at a distance of ten meters. I do not think anyone can do quite as well as that theoretical best possible, but if we had mouse-sized eyes we would see much less well. With their small eyes, mice could not hope to be able to read the small print at the bottom of an optometrist's chart.

With both eye sockets in the front of the skull, making the eyes facing forward, we can use the binocular effect to judge distance. This depends on the left and right eyes seeing the same object from slightly different angles. The larger the difference between the images on the two retinas, the closer the object must be.

## EAR

The working parts of the ear are hidden away in the skull, immediately behind the jaw joint. [Plate 26 and Figure 6] A bony tube opening here leads to the eardrum, a delicate membrane of about nine millimeters diameter. Behind the eardrum, deeper inside the skull, is the air-filled cavity of the middle ear. Sounds make the eardrum vibrate, and its vibrations are transmitted by the ear ossicles (a chain of three tiny bones) across the middle ear to the sensitive organs of the inner ear. The inner ear is a fluid-filled cavity in the bone, with just two membrane-covered "windows" to the middle ear. The last ear ossicle in the chain rests against one of the windows, so vibrations of the ossicles make the fluid in the inner ear vibrate. The cells that detect these vibrations are in the coiled part of the inner ear known as the cochlea.

This mechanism for hearing may seem unduly elaborate, but it solves a problem. It is very difficult to transmit sound waves from air to water. When sound traveling through air hits a water surface, 99.9% of the sound energy is reflected and only 0.1% (or less, depending on the angle) gets into the water. That means that if the eardrum were placed directly over the window into the inner ear, with no intervening ear ossicles, hardly any sound energy would reach the sensitive cells. What makes it so difficult for sound to pass from air to water is that water is much denser and less compressible than air. A sound traveling through air involves relatively large vibrations of the air molecules and small oscillations of pressure. A sound of the same pitch and energy content in water has much smaller vibrations and much bigger pressures. To transmit a sound effectively from air to water you need to make the vibrations smaller and the pressure fluctuations larger.

That is what the middle ear does. The window into the fluid-filled inner ear, to which the ossicles transmit vibrations, is very much smaller than the eardrum. By concentrating the forces onto a smaller area, the pressures are increased. In addition, the ossicles are a system of levers that increase the forces a little and reduce the amplitude of the vibrations, helping sound transmission. Further help is given by the external ear, which concentrates sound from a large area onto the much smaller area of the eardrum. With all these design features, sound is transmitted very effectively into the inner ear.

Eyes on the front of the head enable us to use the binocular effect to judge distance. Our ears are at the sides, which is better for sensing the direction of a sound. Sound from one side of the body reaches the near ear before the far one, but this seems unlikely to be useful for sensing direction because the delay is too short. Sound travels through air at a speed of 330 meters per second, so takes only a fraction of a millisecond to travel the width of the head. A more useful indicator of sound direction is the difference in sound intensity at the two ears. The sound is less loud at the farther ear, because it is in the sound shadow of the head. This works well only for wavelengths that are fairly short compared to the diameter of the head—that is for high-pitched sounds. It does not, however, make it as much easier as you might suppose to detect the direction of higher- pitched sounds, because of confusing echoes. Higher-pitched sounds are more strongly reflected from surrounding objects.

Two ears are not essential for judging sound direction; people can still judge it quite well with one ear blocked. This seems to be due to the directional properties of the external ear.

## INNER EAR

As well as the cochlea, the inner ear contains the organs of balance. The three semicircular canals, arranged roughly at right angles to each other, tunnel through the surrounding bone. They detect rotations of the head. When you rotate a cup of coffee, the coffee does not keep pace with the cup but lags behind, a consequence of its inertia. Similarly, when the head turns, inertia makes the fluid in the canals lag behind. Sense organs in the canals detect the movement of the fluid relative to the canal wall.

The system does not work with quite the elegant simplicity that that may suggest. It used to be assumed that fluid was made to flow in the horizontal semicircular canal

*Plate 22 (opposite) The inside of the skull roof is smooth, apart from the grooves for blood vessels. (Courtesy Maxilla & Mandible.)*

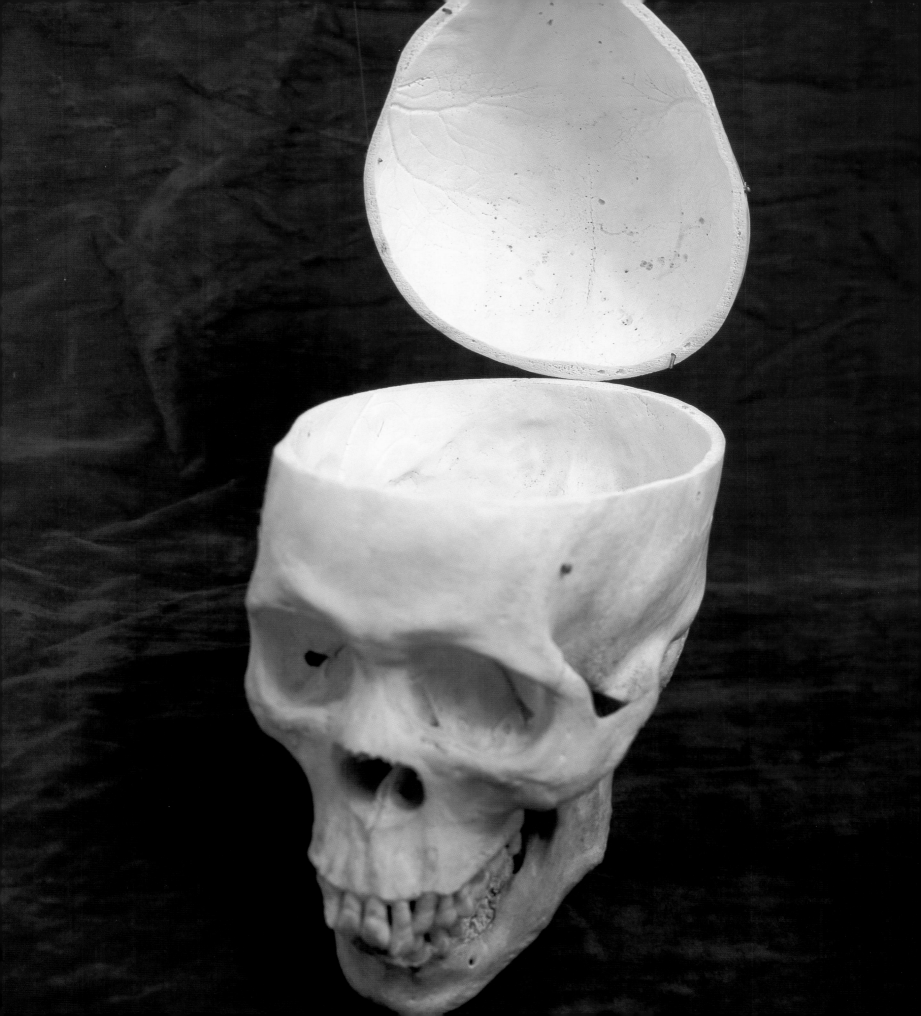

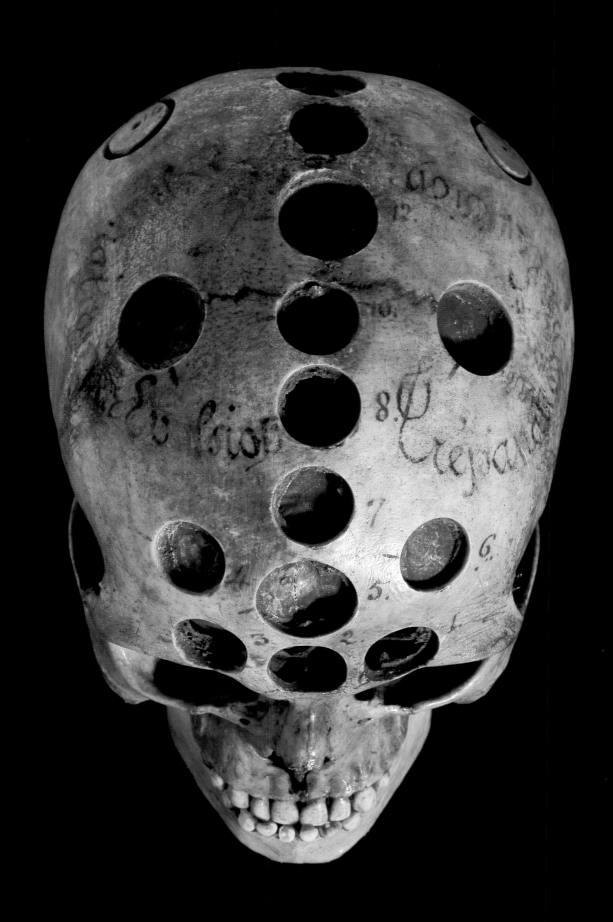

only by rotations in the horizontal plane, and that the two vertical canals were each affected only by rotations in their own planes. It is now known from experiments with glass models and from mathematical analysis that rotation in the plane of one canal sets the fluid moving in the others as well. Happily our brains are able to unscramble the signals from the canals and tell us the direction of any rotation.

The inner ear has two more sense organs in its globular central part. The otoliths are masses of calcium carbonate crystals held together by jelly-like material. They are loosely attached to the inside of the wall of the inner ear, over groups of sense cells that detect any movement of the otolith relative to the wall. The otoliths are much denser than the fluid in the inner ear, so if the head accelerates in any direction the otolith tends to lag behind. One otolith lies horizontally on the floor of the inner ear. Accelerations in horizontal directions make it slide over the underlying sense cells and are detected by them. Separate groups of sense cells detect movement in different horizontal directions, enabling us to distinguish forward or backward accelerations from sideways accelerations. The second otolith organ is on one of the vertical walls of the inner ear, and is sensitive to vertical accelerations. Between them, the two otolith organs can register acceleration in any direction. We are getting information from them when we sense the forward acceleration of a motor vehicle, or the vertical acceleration of an elevator.

The signals from the otolith organs are ambiguous because tilting the head also makes them move over their sense cells. Normally our eyes can tell us whether we are accelerating or tilting, but confusion is possible in darkness or mist, and is thought to have been the cause of some air crashes.

## NOSE

The bones of the skull extend into the bridge of the nose, but the lower parts of the nose are stiffened only by cartilage. Consequently the nose appears on a skull merely as a small flange of bone at the top of a triangular hole. A vertical cut through a skull shows that the nasal cavity extends well back into the skull, under the brain. [Plate 27] The sense cells that detect odors lie deep within this space. They are connected to the brain by a large number of slender nerves that pass through the holes in a sieve-like bone in the floor of the braincase. [see Plate 24, near the top of the photograph] I see no special advantage in this arrangement; a few larger nerves could have done the same job.

Within the nasal cavity are several exceedingly delicate curved sheets of bone, the nasal conchae or turbinals. [Plate 28] The sense cells are on the hindmost of these. The other turbinals have a function connected with breathing that is very important to some mammals, but less effective in humans. They are covered by a layer of tissue that

*Plate 23 (opposite) A skull from the estate of Dr. Louis Auzoux, physician to Napoleon III. The holes were cut to show the thickness of the bone at points considered important in phrenology. (Courtesy Henry Galiano.)*

is kept moist by the secretions of mucus glands. These delicate structures conserve heat and water. Unless conditions are most uncomfortably hot, the air around us is much cooler than our body temperature of 37°C. The air that we breathe inevitably gets warmed up to body temperature by the time it reaches the lungs. Warming makes it capable of taking up more water, so water evaporates into it from the moist internal surfaces of the body. If we breathed the air out again at body temperature, we would lose both the water it had taken up, and the heat that was needed to evaporate that water.

The air passes through the narrow gaps between the turbinals, as we breathe in and out through the nose. Cool air coming in from the nostrils and warm air passing out from the lungs set up a temperature gradient in the turbinals, cool at the nostril end and warm further up the nose. Air being breathed in travels up the temperature gradient, getting warmed up and taking up water. When it is breathed out again, it travels down the gradient and is cooled, and water condenses out from it. It leaves the nose saturated with water, but at a temperature that may be well below body temperature. At this lower temperature it contains less water vapor than when it was warmer.

The effect has been investigated by putting microbead thermistors (tiny electrical thermometers) into the nostrils of people and animals. People usually breathe air out at about 33°C, only a few degrees below our body temperature of 37°C. Kangaroo rats from North American deserts do much better. Their exhaled air is slightly cooler than the atmosphere, for example 18°C in an experiment in a laboratory where the air around them was at 22°C. Consequently, they lose very little water vapor in their breath. This is one of the features that make them far better able than people, to survive in deserts.

Hidden away in the bones of the face are a number of air-filled cavities, the paranasal sinuses. There are sinuses on either side of the nose and immediately above it, all connected to the nasal cavity. Unfortunately, infections spread easily into them from the nasal cavity. We are generally not aware of our sinuses until they cause trouble.

## CRANIAL NERVES

The brain sends out many nerves, known as cranial nerves, through holes in the skull. Some of the holes are labeled in Figure 6. The optic nerve to the eye emerges through a hole in the bone of the eye socket. [Plate 29] An L-shaped slit below it is the passage for three nerves that between them serve the six eye muscles, and also for two of the three branches of the very large nerve that is called the trigeminal. One of these two branches goes to the scalp and the upper part of the face. The other burrows into the bone again, emerges through a hole below the eye, and serves the upper jaw. The third branch emerges from the underside of the skull and

PLATE 24
*The complex shape of the floor of the braincase*
*fits the underside of the brain.*
*(Courtesy Maxilla & Mandible.)*

innervates the lower jaw and the jaw muscles. The remaining cranial nerves all emerge from the underside of the skull. The facial nerve goes to the parts of the face that are not served by the trigeminal. The glossopharyngeal nerve serves organs within the mouth. It shares a hole in the skull with the vagus, which sends branches far down into the torso to help control the heart and guts. Two more cranial nerves go to the neck muscles and tongue.

The cranial nerves have played a very important part in the development of theories of the evolution of the head. We humans are members of a great group of animals known as the vertebrates: the fishes, amphibians, reptiles, birds and mammals. The name of the group implies that its distinctive characteristic is possession of a backbone, but hagfishes (which are very primitive fish) have no backbone. Instead they have a notochord, a flexible fluid-filled rod running the length of the body, that performs the same mechanical function as a backbone but has no vertebrae associated with it. The most characteristic features of the vertebrates are not the vertebrae, but the brain and skull.

## ORIGINS OF THE SKULL

Many invertebrate animals (for example, earthworms and caterpillars) have bodies divided into a large number of similar segments. Vertebrate animals are also segmented. This is most clearly seen in fish, but is partly obscured in mammals such as ourselves. The head of a fish is not obviously segmented, but the swimming muscles are divided into W-shaped blocks called myomeres, which separate easily when the fish is cooked. There is a vertebra for every pair of these blocks (one block on each side of the body), and a pair of nerves emerges between each vertebra and the next. Thus the body of the fish, behind the head, is a series of segments, each with a pair of myomeres, a vertebra and a pair of spinal nerves. The segmental structure of the muscles of the torso is less obvious in mammals such as ourselves, and we have more variation in structure between successive vertebrae, but the basic segmental structure remains. Human embryos have myomeres, just like the myomeres of fish embryos.

It has been suggested that segments that were formerly part of the trunk may have been incorporated in the heads of vertebrates. In 1791, Johann Goethe, the great German poet, picked up a sheep's skull in the Jewish cemetery in Venice. He was an anatomist as well as a poet, and it occurred to him that the skull was a series of modified vertebrae. He did not publish the idea until 1820, by which time the German nature philosopher Lorenz Oken had independently formulated and published the same idea. Their theory did not imply evolution (Charles Darwin's *The Origin of Species* did not appear until 1859), merely that the skull was in some fundamental way equivalent to a series of vertebrae. It was discredited by the English biologist Thomas Huxley, who

*Plate 25 (opposite) The spinal cord joins the brain through the foramen magnum, the big hole in the underside of the skull. (Courtesy Maxilla & Mandible.)*

*Plate 26 The tube that leads to the eardrum opens close behind the jaw joint. (Courtesy Maxilla & Mandible.)*

was to become Darwin's most vocal supporter, in 1858. That theory was abandoned, but a succession of later anatomists formulated a less extreme theory. This new theory was presented in its most developed form in 1930 in Edwin S. Goodrich's *Studies on the Structure and Development of Vertebrates*. This theory will bring us back to cranial nerves, but we need first to look at spinal nerves.

Each spinal nerve leaves the spinal cord in two parts, the dorsal root (behind) and the ventral root (in front). These two roots merge to form the main trunk of the nerve, which then splits again to form branches to various organs. Similarly, as you go up a tree, you find roots merging to form the trunk, which then splits again into branches. The dorsal root of a spinal nerve consists largely of sensory nerve fibers, and bears a ganglion (a swelling filled with nerve cells) close to its base. The ventral root consists of motor fibers to the myomeres. Some cranial nerves (the trigeminal, facial, glossopharyngeal and vagus) resemble dorsal roots. They have ganglia, and many sensory fibers. Others (the nerves of the eye muscles) are more

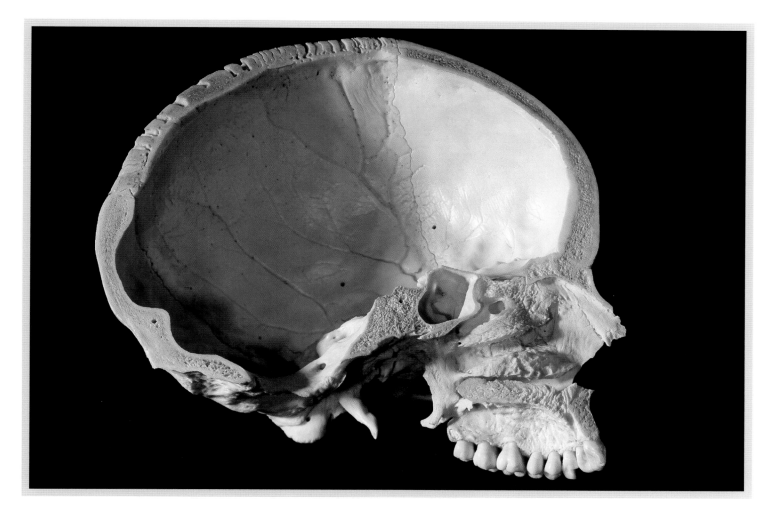

like ventral roots. There is no joining of dorsal and ventral roots among the cranial nerves, but Goodrich and like-minded anatomists argued that these cranial nerves were indeed dorsal and ventral roots of segments that had become incorporated into the head.

When I was a student in the 1950s, this theory was taught in detail. It was considered essential knowledge for every zoologist. It became unfashionable, and featured in very few biology courses by the end of the century. Interest in it has been re-awakened by recent research, which has given it very strong support.

For a short time during the development of vertebrate embryos, the hind part of the brain has a series of constrictions that make it look rather like a short piece of the segmented body of an earthworm. The segment-like divisions of the brain are known as rhombomeres. They have been studied in zebra fish, chick and mouse embryos, and are presumably formed also in humans.

*Plate 27 A vertical section through a skull shows the nasal cavity below the braincase. (Courtesy Maxilla & Mandible.)*

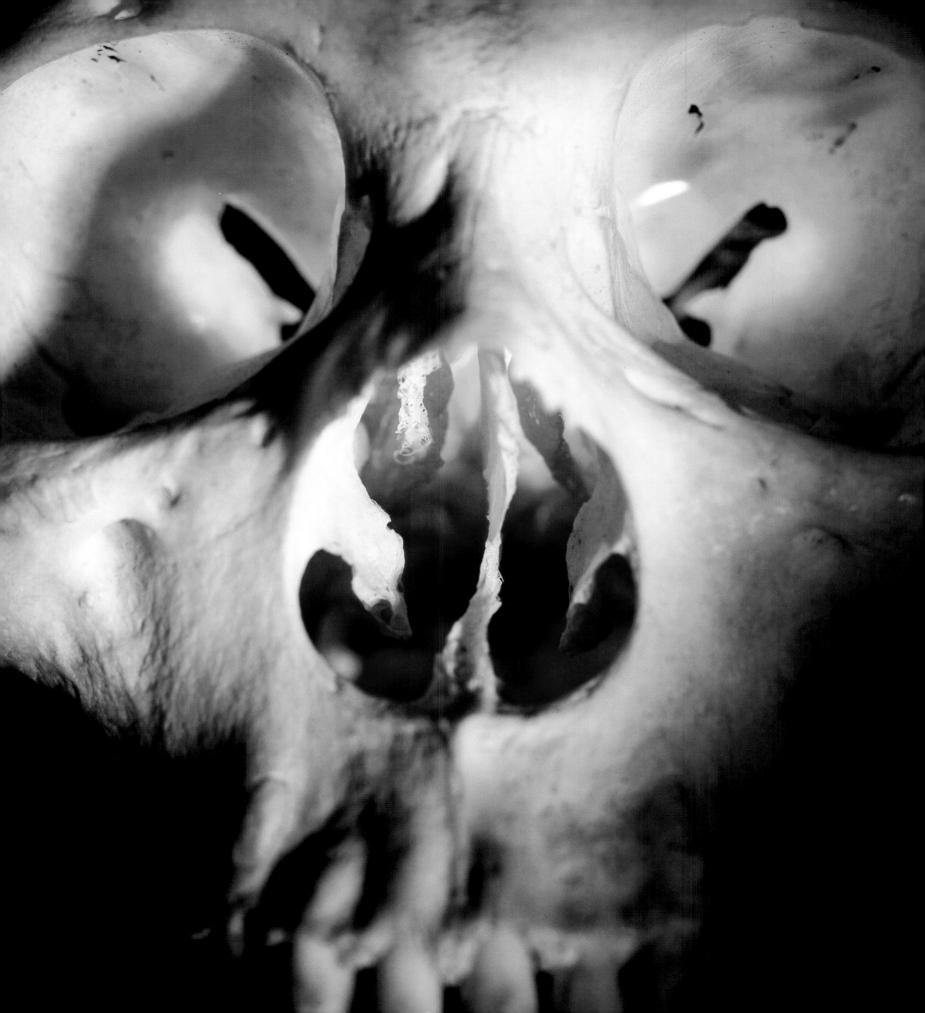

There are seven or eight of these rhombomeres. They are not all alike. Separated cells from even-numbered rhombomeres will adhere to cells from other even-numbered rhombomeres, but not to cells from odd-numbered rhombomeres. In mixtures of separated cells, cells from odd- and even-numbered rhombomeres sort themselves out and form separate clumps. Thus two rhombomeres (one odd and one even) make a unit in the repeat pattern of the developing hind brain.

Each rhombomere pair gives rise to a pair of dorsal root nerves. Those from the first rhombomere pair become the trigeminal nerves; those from the second rhombomere pair become the facial nerves; and those from the third and fourth pairs become the glossopharyngeal and vagus nerves. It seems that each rhombomere pair corresponds to a segment in the spinal cord.

The implication seems to be that in the course of evolution of the first vertebrates, four segments from the front end of the trunk became incorporated in the head. The eye muscles seem to be all that remains of the myomeres of the first three of these segments, and the three nerves that serve them are their ventral roots. The myomeres of the fourth segment seem to have disappeared entirely; the inner ear occupies the space where you might expect to find them. Thus the skull did not evolve from vertebrae. The relationship between its structure, and the structure of the vertebrae, is a good deal subtler than that.

The amphioxus is a fish-like animal, a few centimeters long, which lives buried in sand offshore. It is obviously a primitive relative of the vertebrates, but it has no vertebrae, no skull and no proper brain. Study of the genes that control segmentation have shown that its front four segments correspond to the four rhombomere pairs of vertebrates. It has a short extension of the spinal cord in front of these segments. This is presumably the structure that evolved in an amphioxus-like ancestor of the vertebrates to form most of the brain. Development of the human brain from such modest beginnings was a massive evolutionary advance.

## PALATE AND JAWS

The palate is a plate of bone that forms the floor of the nasal cavity and the roof of the mouth cavity. [see Plate 27] It protects the delicate turbinals from damage by food in the mouth, and it forces the air to pass over all the turbinals before it can reach the mouth cavity. The opening of the nasal cavity into the mouth is well back in the mouth, close to the opening of the windpipe at the top of the throat. Thus the air by-passes the mouth cavity, making it possible for us to breathe with our mouths full. This is very helpful, when food needs prolonged chewing. Chewing food when your nose is blocked by a heavy cold can be a most uncomfortable experience. Other mammals have palates like ours, but the nostrils of most reptiles and of birds open

*Plate 28 (opposite) Parts of the nasal conchae can be seen in this view into the nose. (Courtesy Maxilla & Mandible.)*

directly into the front of the mouth. They do not need the by-pass arrangement because they do not chew their food.

The upper jaw and palate develop before birth, from bones that grow inward from either side of the head. Occasionally the two sides fail to meet, causing the defect known as cleft lip and palate. [Plate 30] This can be corrected by surgery.

Babies have two lower jaw bones, one on each side, joined by collagen fibers at the chin. [see Plate 13] The two bones fuse together within three years, so older children and adults have just one lower jawbone. [Plate 31] Rounded knobs (called condyles) at the back of the lower jaw fit into hollows in the underside of the skull to form the jaw joints. Looking at the bones you might think that the joints worked as simple hinges, but the ligaments that hold them together are loose enough to allow more varied movement. As well as opening and closing our mouths (the hinge action) we can slide our jaws forward and back a little. Moving the jaw forward dislocates the joint, sliding the condyles forward out of the hollow that they normally rest in. We can also move our jaws from side to side. All these movements play their part in eating, as we will see.

There are pointed projections from the top of the lower jaw, in front of the jaw joints. These are the coronoid processes. When the mouth is closed, each coronoid process lies between the main body of the skull and a bar of bone, called the zygomatic arch that runs from behind the eye socket to above the bony tube that leads into the ear.

The areas of attachment for the jaw muscles are shown in Figure 6. The strongest of them is the masseter muscle, which runs upward and forward from the outer surface of the lower jaw to the zygomatic arch. Next in order of strength is the temporalis muscle, which fans out from the coronoid process and attaches to the side of the braincase. A slight ridge, which runs in a curve high up on the side of the braincase, marks the upper edge of the temporalis. The medial pterygoid muscle is on the inner side of the jaw. All these muscles close the mouth, and are used in biting and chewing. The lateral pterygoid muscle, also on the inner side of the jaw, pulls the whole jaw forwards.

The digastric muscle opens the mouth. It runs from an attachment on the jaw close behind the chin, to an attachment on the skull close behind the jaw joint. A direct path between those attachments would pass through the mouth cavity, but the digastric muscle follows a V-shaped path, down into the throat, round a loop of collagen fibers attached to the hyoid bone, and up again. (The hyoid bone, Plate 32, is a remnant of the gill skeleton of our fish ancestors.) The loop of fibers functions like a pulley wheel. The name "digastric" means two-bellied. The muscle has one fleshy belly running down from the skull, then a length of tendon running through the hyoid loop, and then a second fleshy belly running up to the chin.

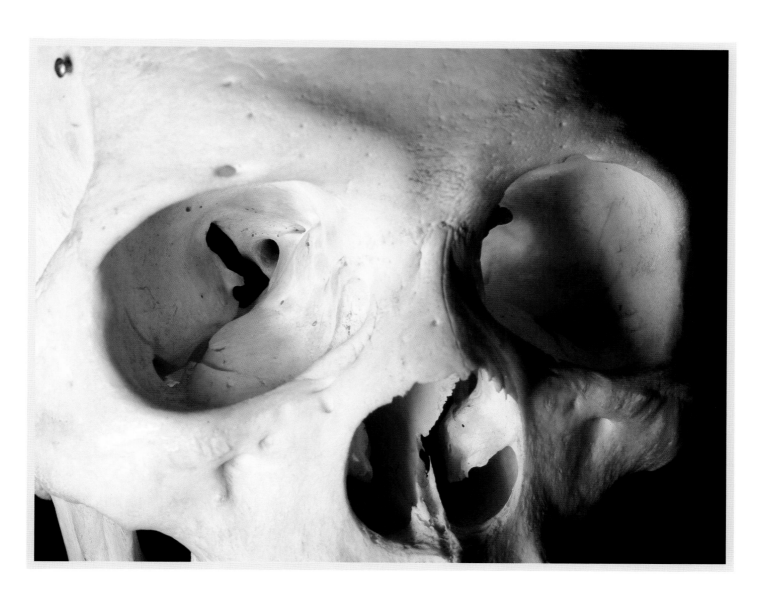

PLATE 29

*The optic nerve, and nerves to the eye muscles and*
*face, passed through the holes in the wall of the eye*
*socket. (Courtesy Maxilla & Mandible.)*

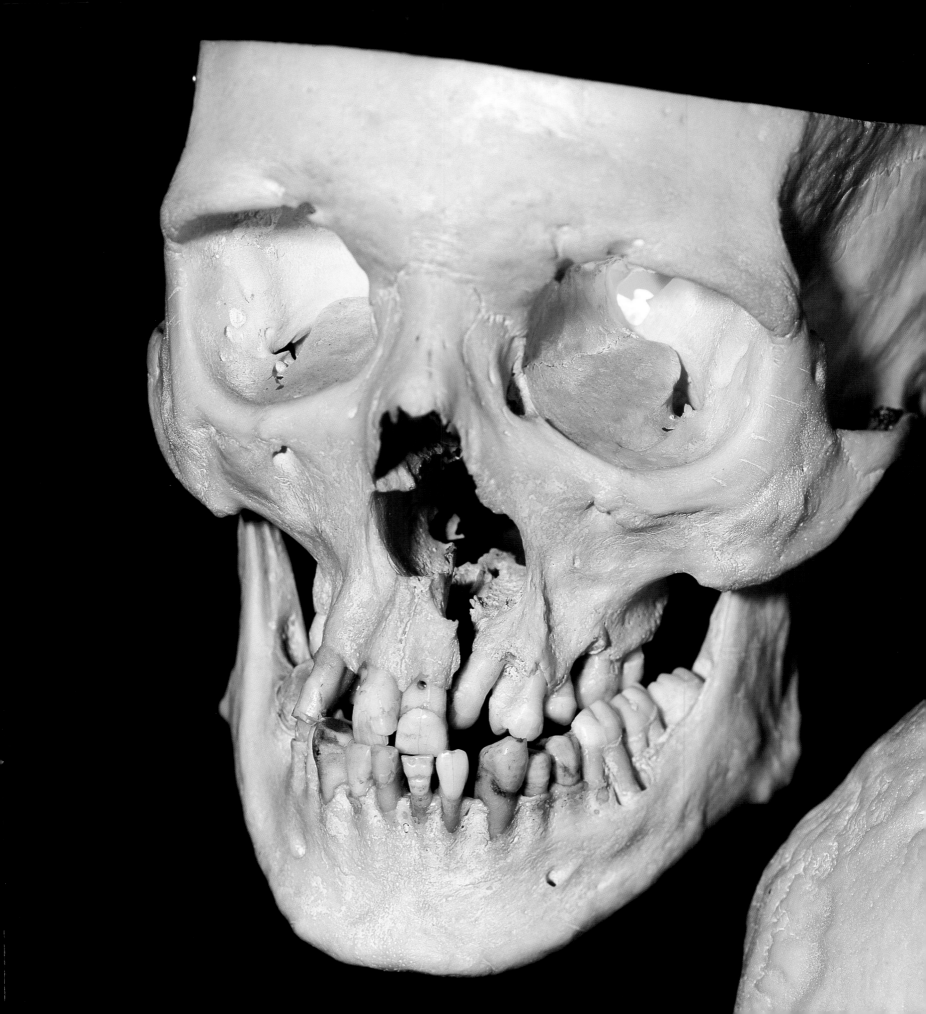

## TOOTH STRUCTURE

We have 32 teeth, of four different kinds. [Plate 33] At the front are the square, sharp-edged incisors, two on each side in each jaw. They are good for biting off pieces of food. Next are the pointed canines, one on each side in each jaw. In carnivores, the canines are the fangs that are used to kill prey and tear flesh, but our canines are much too short for those purposes. I cannot think of any task, for which I would use my canines in preference to my other teeth. Their pointed shape is a relic of our ancestry. The canines are followed by two premolar teeth on each side in each jaw. Whereas the canines have a single pointed cusp, the premolars each have two cusps side by side. Finally, behind the premolars, are the molars. There are three of them on each side of each jaw. The molars have four cusps, placed as the corners of a square. [Plate 34] We use our premolars and molars for chewing.

All the teeth have roots, which are firmly fixed by collagen fibers in their sockets in the jaws. The incisors and canines have one root each; most of the premolars have one but the first upper ones have two; and the molars have two or three roots. All the teeth are hollow, with a pulp cavity filled with soft tissue. Blood vessels and nerves enter the pulp cavity through openings at the tips of the roots.

When the mouth is closed in its normal resting position, the crowns of the upper and lower molars are in contact, but the lower incisors lie behind the upper ones. To make the upper and lower incisors meet edge to edge (for example, to take a bite from a cookie) we have to use the lateral pterygoid muscle to move the jaw forward, disarticulating the jaw joint. With the jaw in this position, the upper and lower molars are separated. We have to move our jaws forward to use our incisors and back to use our molars.

The upper and lower molars fit very neatly together, with the cusps of the upper molars in the hollows of the lower ones, and vice versa. [Figure 7] We feel uncomfortable and cannot chew so well if a dentist, making a filling, gets the fit even slightly wrong.

We can use our molars to crush food or to grind it. If the lower jaw is moved vertically up, directly into the position in which the molars fit closely together, the food is merely crushed. This is fine for cracking brittle foods such as nuts or squeezing the juice out of fruits and other soft foods, but a grinding action is needed to break up tougher food. To achieve this, the lower jaw must be brought up obliquely, so that the crowns of the teeth slide across each other as they approach the close-fitting position. The movements of chewing have been observed by filming the front teeth of Aboriginal Australians, who habitually chew with their lips parted, and by X-ray cinematography of other people. Normal chewing involves elliptical movements of the jaw. If for example the food is being chewed on the left side of the mouth, the lower jaw is moved to the left as it opens and back towards the centre as it closes again. Chewing cattle make similar but more obvious movements.

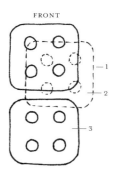

FRONT

*Figure 7 How the upper and lower molars fit together. (1) side, (2) upper molar, (3) lower molar*

*Plate 30 (opposite) A skull with a cleft palate. (Courtesy Henry Galiano.)*

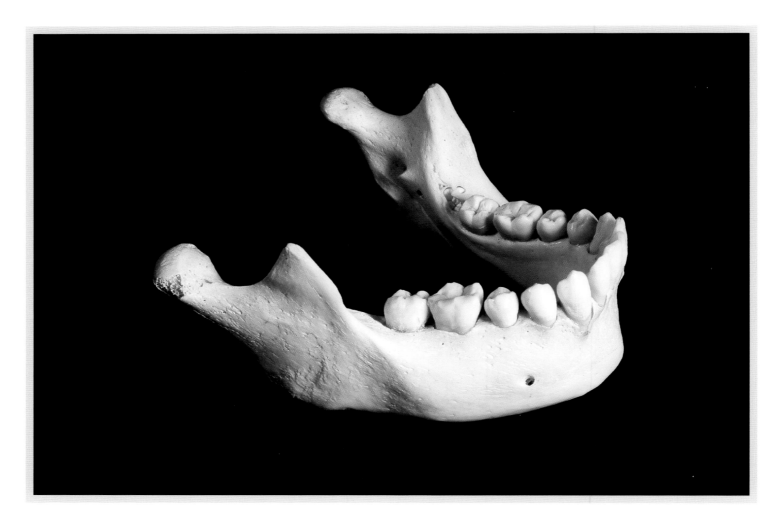

*Plate 31 A child's lower jaw. The four-cusped milk molars would eventually have been replaced by two-cusped adult premolars, if the child had survived. (Courtesy Maxilla & Mandible.)*

Teeth consist mainly of dentine (ivory), which is very similar in composition to bone. The main difference is that bone has living cells scattered all through it, but dentine has all its cells in the soft tissue of the pulp cavity. The crowns of the teeth are covered by a layer of enamel, 2.5 mm thick in its thickest parts. Enamel is harder than bone and more resistant to wear, but also more brittle. Once they have erupted, our adult teeth have to last for the rest of our lives, so wear resistance is very important. They also survive well after death, better than any other part of the body. Teeth are the commonest fossils of humans and other vertebrates.

## ENGINEERING VS. EVOLUTION

Much of this chapter has been about the functions of different parts of the skull, but I have referred at several points to our ancestry. For example, our canine teeth are the rudiments of our ancestors' fangs. The cranial nerves, which emerge through holes in the skull, reflect the segmented structure of the earliest

vertebrates. The pattern of bones in the roofs of our skulls has not been designed afresh by evolution, to suit the particular needs of the human race, but has changed only a little from the pattern in our amphibian ancestors of 350 million years ago.

There is a fundamental difference between the processes of evolution and of engineering design. An engineer who has designed a boat, and wants next to design a car, goes (literally) back to the drawing board, and builds up the new design from scratch. Evolution, however, starts from an existing design and alters it progressively by a series of small changes over many generations. The final product (for example, an amphibian) may be very different from its ancestor (for example, a fish), but every stage in the evolutionary sequence must be an effective design, capable of holding its own in a competitive world. As a result, many of the features of a species may not be ideal for its way of life, but may be (more or less) the best that evolution could do, starting from that species' ancestors.

*Plate 32 The hyoid bone in place in the throat. (Courtesy Division of Anthropology, American Museum of Natural History.)*

*Plate 33 (following pages) A complete set of teeth. (Courtesy J. G Sawyer.)*

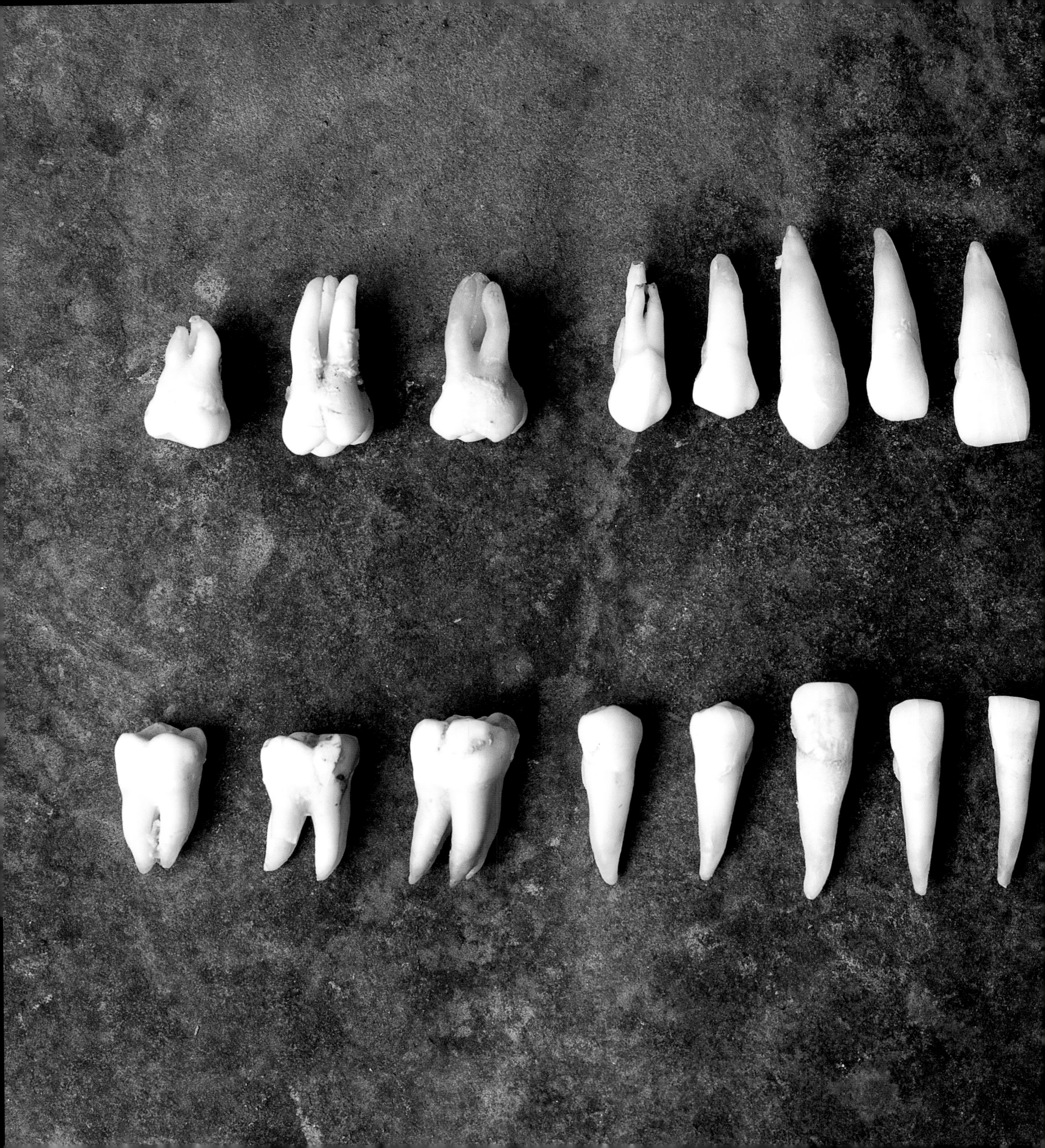

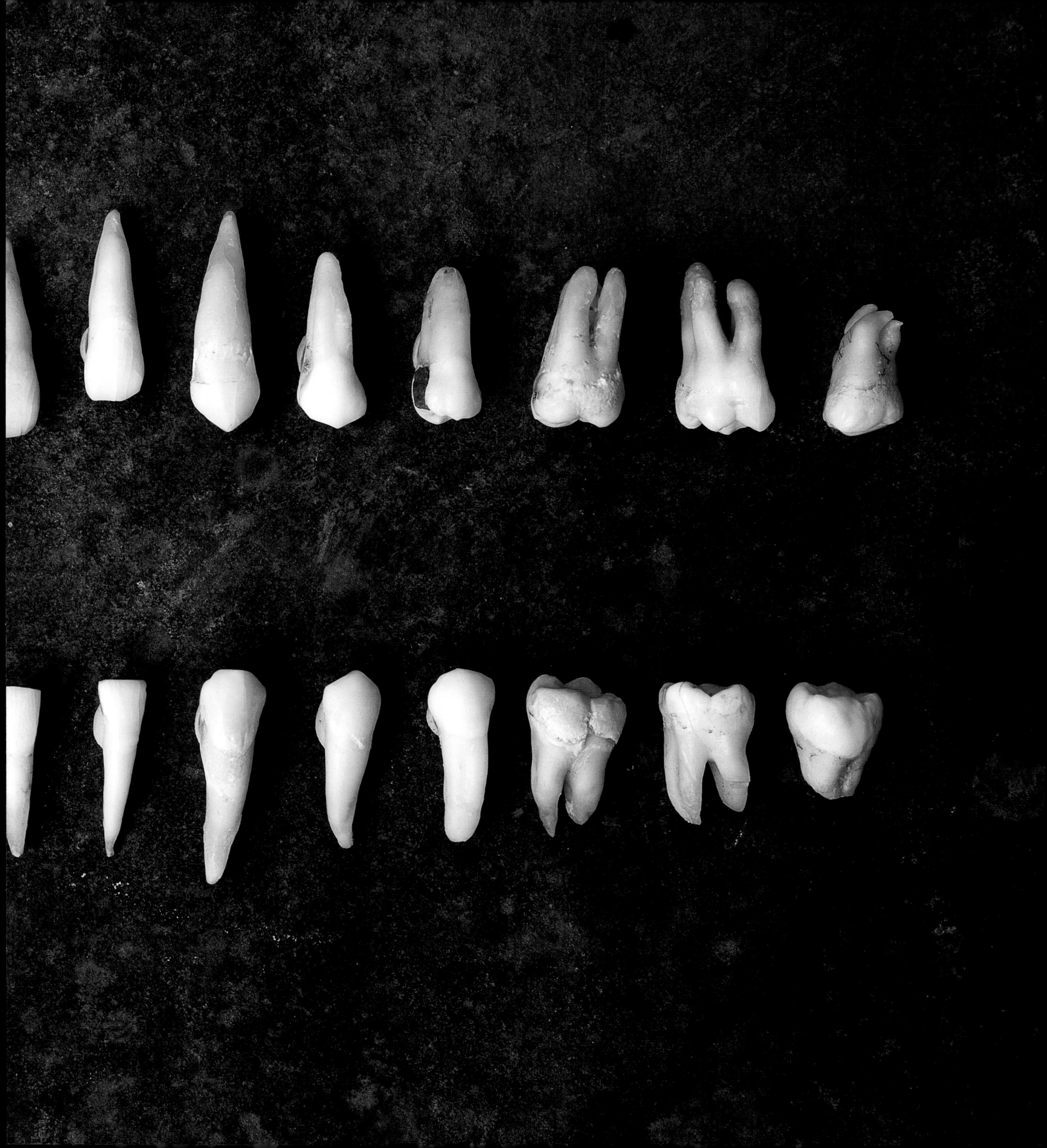

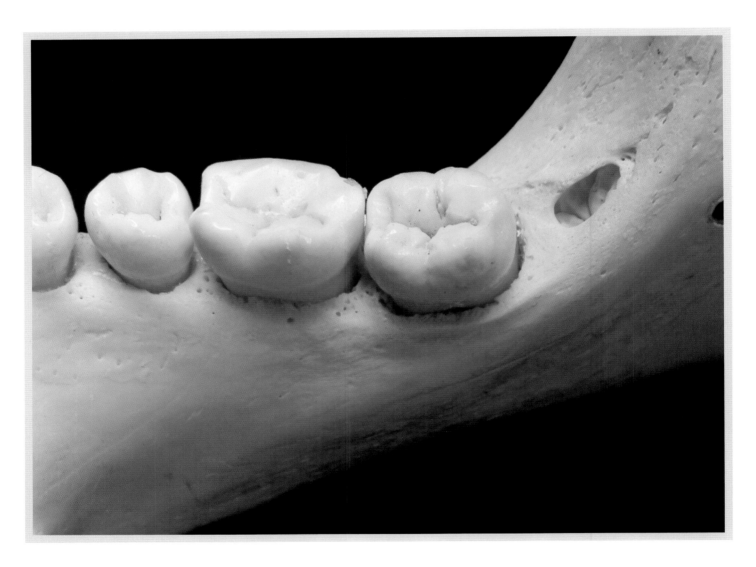

PLATE 34
*Premolars (the smaller teeth) and molars (the large ones) in an*
*adult jaw. The third molar (the "wisdom tooth") has not erupted.*
*(Courtesy Maxilla & Mandible.)*

I like to explain this point by comparing an evolving animal to a walker in a mountainous landscape. If the walker always chooses an uphill path, he or she may reach the highest peak in the range. More probably, they will finish at the top of some subsidiary peak from which they cannot climb higher without first descending. Their final height will be different, depending on the point from which they started. Evolution by natural selection (the survival of the fittest) may be incapable of taking an animal from a subsidiary peak of fitness to a higher peak, because that would require a descent into a less fit state. The pattern of bones in human skull roofs is not necessarily the best possible for us, but is a pattern that could be evolved from our ancestors.

A recurring theme of this book will be that the design of the skeleton reflects ancestry as well as function.

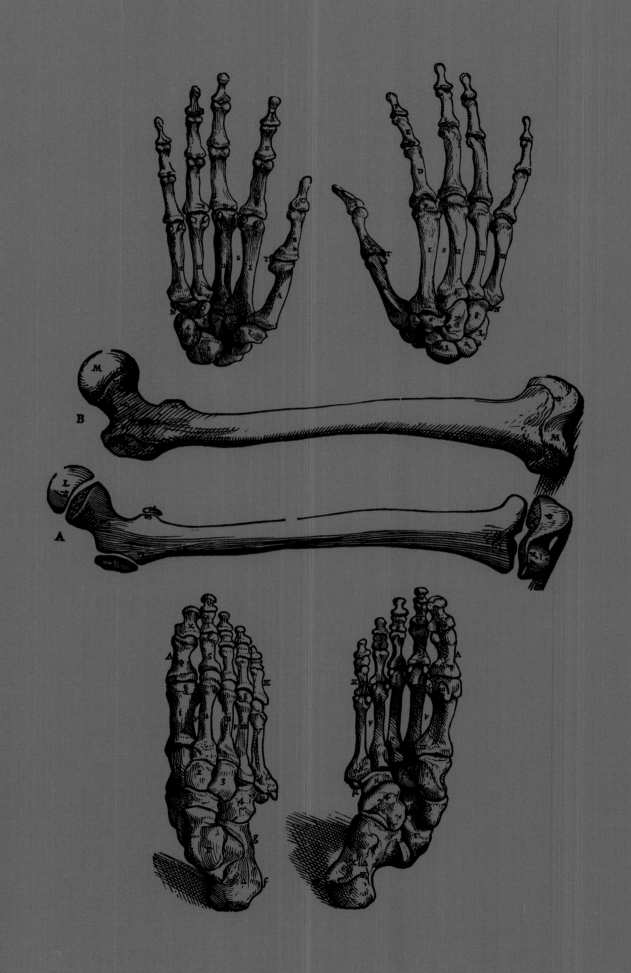

# 3

## ARMS & LEGS

From the skull we move to the very different bones of our arms and legs. The skull is built from plates of bone, but our principal limb bones are rods. The largest of them are the humerus, the bone of the upper arm [Plate 35] and the femur, the thigh bone. Ask someone to draw a bone, and they will probably draw a long, straight shaft with a knob at each end, like the crossed bones of the pirate flag. Of all the bones in the human body, the humerus and the femur are the ones that fit that stereotype best. Some old tombstones are carved in sufficient detail to show that the bones of the skull and cross bones are intended to be femurs, but we need not notice the differences between the humerus and the femur until later in this chapter.

### NATURE OF BONE

"Bone" is a word with two main meanings. It can be used to refer to one of the pieces of the skeleton (such as the humerus), or the material from which those pieces are built. Bones (the structures) consist of bone (the material). A saw cut across the shaft of the humerus [Plate 36] reveals the bone as a whitish, apparently homogenous material, similar in appearance to the plastics that are often used for making handles for table knives. The resemblance is no coincidence. Knife handles were traditionally made of ivory which, as we have seen, is very similar in composition to bone. The plastics were developed to imitate ivory.

The homogenous appearance is deceptive, because bone is actually a blend of two materials, the mineral hydroxyapatite and the protein collagen. Hydroxyapatite is a form of calcium phosphate, also found in some rocks. It makes up about 70% of the dry weight of bone, and is present as crystals so small that an electron microscope is needed to make them visible. The collagen, as long fibers, makes up most of the remaining 30%. The crystals are firmly bonded to the fibers. The dentine and enamel of teeth are also blends of hydroxyapatite and protein, but enamel is harder than bone or dentine because it has more hydroxyapatite (95%).

If bones were made of wet collagen, without any hydroxyapatite, they would be flexible like rubber. If they were made of hydroxyapatite without collagen they would be brittle like glass or china. Neither material would be satisfactory alone, but the combination of the two is tough and stiff, just what is needed to build the framework of the body. The trick of combining two materials to do a job that neither could do alone is well known to engineers. Fiberglass, glass fibers embedded in a plastic resin, is one of the great successes of this technique. A rowing boat made of the plastic alone would be intolerably flexible, and a glass boat would be dangerously brittle, but fiberglass makes excellent boats. Fiberglass was the first widely used man-made composite for applications like these, but composites made with other fibers such as carbon and silicon carbide are now used in racing shells, tennis rackets and other sports equipment. The tires of motor vehicles are made of rubber that by itself would be soft and relatively weak, stiffened and strengthened by incorporating particles of soot.

The trick of combining two or more materials to make a skeleton has also been used repeatedly in Nature. Snail shells consist of tiny crystals of calcium carbonate (the main constituent of limestone rocks) bound together by a little protein. Beetles and other insects are encased in a composite of chitin fibers embedded in protein. Wood is a composite of cellulose and lignin. Both chitin and cellulose are polysaccharides, compounds built from chains of sugar molecules. Lignin is a very complex organic chemical whose structure does not seem to be fully understood.

Bone is a little stronger and stiffer than hardwoods such as oak, but it is also denser than wood because of its mineral content. Weight for weight, bone and timber are about equally strong. Wood splits along the grain more easily than it breaks across the grain, because its fibers are aligned along the grain. Similarly, bone has a grain and is stronger in one direction than in others. In long bones such as the humerus, the grain runs along the length of the shaft, just as in trees the grain is aligned with the branch. Bone, however, does not split along the grain as easily as wood, because the directions of its fibers are more varied.

## BONE STRUCTURE

In Chapter 1 I described the osteons, the cylindrical units from which much bone is built. [see Figure 2] The collagen fibers of the bone run helically (like a spiral staircase) along the osteons. In each osteon, alternate layers have left-handed and right-handed

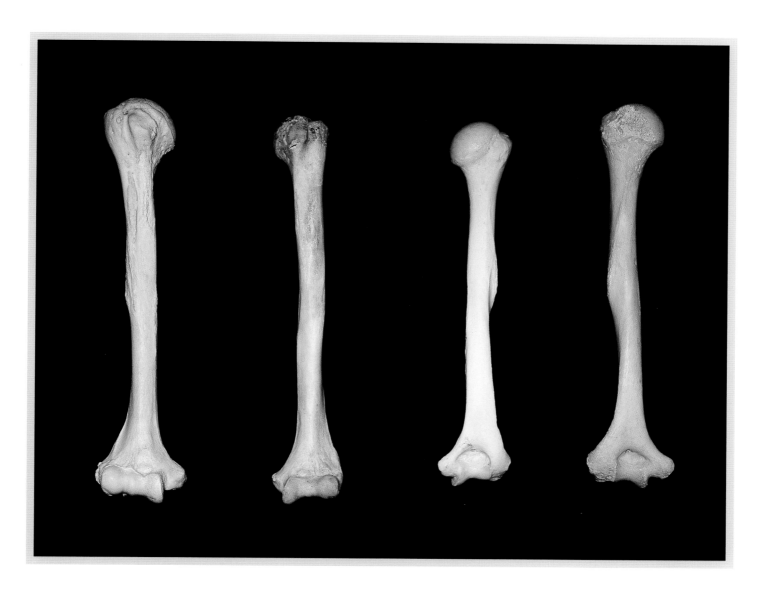

PLATE 35
*Humeri (upper arm bones).*
*(Courtesy Maxilla & Mandible.)*

helices. This arrangement makes the osteons less likely to split. The same principle is used in plywood, which has the grain running in different directions in successive layers. Engineers try to arrange the fibers in materials such as fiberglass in the best directions to take the stresses, but the natural process that builds bones does this far more subtly.

The picture of the humerus cut across [see Plate 36] shows that this bone, like many other limb bones, is hollow. As a hollow tube it can be lighter than a solid bone would have to be, to be equally stiff and strong. Lightness is an advantage, because heavy limbs would be unwieldy.

A little explanation is needed here. If you want to break a stick, you will succeed much more easily if you bend it (by exerting forces at right angles to its length) than you would do by pulling or pushing along its length. Long thin structures, including sticks and girders as well as bones, are most easily broken by bending. Tubes resist bending forces particularly well. The reason is that, when you bend a bar, you stretch one side of it and compress the other. The outer layers of the bar are stretched or compressed most, and provide most of the resistance to bending. If the core of the bar is removed, leaving a hollow tube, the bar is made lighter with hardly any loss of bending strength. This is another example of the basic principle we saw applied in the sandwich structure of the skull bones. Scaffolding poles are long slender structures that need to be strong but not too heavy, so are made as hollow tubes instead of solid rods. Similarly, bicycle frames are made from hollow tubes. A bicycle built from solid steel rods would have to be inconveniently heavy.

Scaffolding poles and bicycle frames have nothing but air inside them. In many birds, the humerus is an air-filled tube. In us and other mammals, however, the humerus and other limb bones are filled with fatty marrow. This seems like a design fault, because the marrow increases the weight of the bone without adding to its strength. Marrow is much less dense than bone, and the marrow- filled humerus is lighter than a solid bone of equal strength would be, but the advantage of the tubular structure is less than if the bone were filled with air.

There are two kinds of marrow, red and yellow. The red marrow has the very important function of producing blood cells, both the red cells that carry oxygen around the body and the white ones that attack bacteria and other threats to health. The yellow marrow consists largely of fat, which seems entirely useless. The body's other fat deposits, under the skin and among the muscles and abdominal organs, are valuable food reserves. Starving animals use these reserves to sustain themselves, but leave the yellow marrow untouched. Children have a lot of red marrow, but most of the marrow in adults is the yellow kind. We seem to carry a lot of unnecessary weight around, inside our hollow bones.

*Plate 36 (opposite) A humerus cut through, showing that the bone is hollow. (Courtesy Maxilla & Mandible.)*

A cut lengthways along either the humerus or the femur [Plate 37] shows the structure changing towards the ends of the bone. For most of the length of the shaft the bone is a tube with a wall of compact bone. Close to the ends, it is no longer tubular, but is filled with spongy bone. The sponge-filled ends of the femur form joints with other bones at the hip and knee. As the joints move, different parts of the ends of the bones press on each other. The spongy bone provides support for all the parts of the joint surface that may come under load. Spongy bone is of course lighter than compact bone would be, and it is strong enough for this job. A close look at the spongy bone in the upper end of the femur shows that the slender bars of the spongy network are lined up in particular directions. These directions correspond to the stresses that would act on each part of the living bone.

## JOINTS

The ends of the humerus need to be shaped appropriately, to allow the joints to move. The shoulder end has a knob with a hemispherical surface that fits into a hollow in the scapula (shoulder blade) to form a ball and socket joint. [Plate 38] This kind of joint can be turned in any direction. The rear-view mirror in my car is mounted on a ball and socket joint, enabling me to angle it in whatever way suits me best. I have a reading lamp with a ball and socket joint that allows me to point it in any direction.

Ball and socket joints allow great freedom of movement, but this makes it tricky to fasten the moving parts together. The designer of the mirror in my car has solved the problem by making the socket a little more than a hemisphere, so that the ball is trapped in it and cannot escape. The socket of the shoulder, however, is much less than a hemisphere. Its hollow is merely saucershaped, and does not grasp or enclose the head of the humerus. Other joints are held together by ligaments, as we will see, but there is no way in which the ligaments of the shoulder could be arranged so as to hold the joint firmly together, without restricting its freedom of movement. The shoulder is held together effectively only by the muscles that connect the shoulder blade to the humerus. The muscles can be stretched and the shoulder dislocated, for example, by a fall on the outstretched hand. Dislocation of the shoulder is a common accident in sports. Fortunately, the head of the humerus can be moved back into its socket by manipulation without surgery.

The other end of the humerus has a quite different joint. Instead of a hemispherical joint surface it has a spool-shaped one that fits neatly into a matching hollow in the ulna, one of the two bones of the forearm. [Plate 39] This arrangement makes a hinge joint, allowing the ulna to rotate around the central axis of the spool. Similarly, the hinges of a door allow rotation around a single axis, and no other movement. The elbow is held together by ligaments that consist mainly of collagen fibers. [Figure 8] These radiate out from either side of the humerus, from attachments

Plate 37 *The upper end of a femur cut lengthwise, to show the spongy bone inside it. (Courtesy Maxilla & Mandible.)*

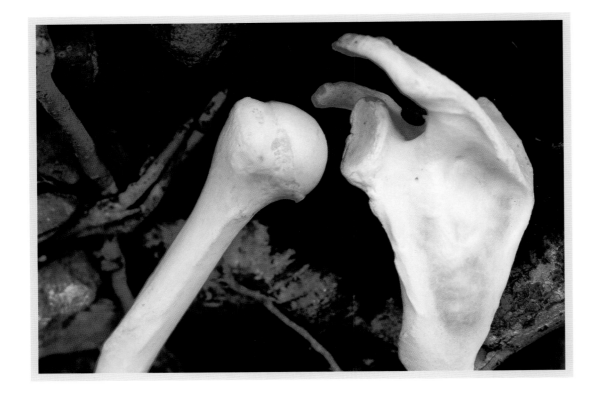

Plate 38 The rounded upper end of the humerus fits into a hollow in the shoulder blade, forming a ball and socket joint. (Courtesy Maxilla & Mandible.)

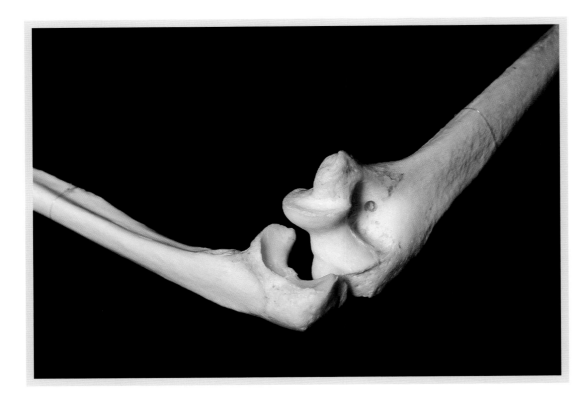

Plate 39 The spool-shaped lower end of the humerus fits into a matching hollow in the ulna, forming a hinge joint. (Courtesy Maxilla & Mandible.)

close to the axis of the joint, to attachment points along the rim of the hollow in the ulna. They protect the elbow from dislocation while allowing it to bend and extend freely.

Here is a simple experiment you can try, to confirm that the joint between the humerus and ulna is a hinge. Sit with your right forearm horizontal, resting your elbow on the arm of your chair. Hold the upper arm with the fingers of your left hand, so that you can feel the humerus inside it. You will find that you can bend and extend your elbow without any movement of the humerus. Now swing your forearm from side to side. This may seem like an elbow movement, but you will feel your humerus rotating; the movement in this case is actually at the shoulder joint.

## "DEGREE OF FREEDOM"

Engineers find the concept of "degree of freedom" useful, in their discussions of joints. The position of a hinge joint can be described by just one number. For example, I can tell you how far open a door is either by stating its angle to the wall, or by telling you the distance between the edge of the door and the jamb. One of these numbers is enough; there is no need to tell you both. Accordingly, the joint between the door and the wall (or any other hinge joint) is said to allow one degree of freedom of movement. A ball and socket joint can rotate about any axis through its centre, but three numbers are enough to describe its position. For example, if you are standing with your arm hanging down by your side, you can swing your arm forward and back; or you can swing it out to the side; or you can rotate the whole arm about its long axis. All these things are done by movement at the shoulder. Any shoulder movement can be described as a combination of these three. Accordingly, the shoulder and other ball and socket joints are said to allow three degrees of freedom of movement.

The next joint as we continue down the arm beyond the elbow is another hinge joint. Sitting with your forearm resting on a table, you can turn your hand palm up or palm down. These actions move the radius (the second bone of the forearm) relative to the ulna. In the palm-down position the radius crosses over the ulna [Plate 40], but in the palm-up position they lie parallel to each other [Plate 41]. You can check that this happens by using your other hand to feel the movement of the bones under the skin. The crossing and uncrossing movement may seem complicated, but it is merely the action of a rather odd sort of hinge. The elbow end of the radius is cylindrical, and fits into a hollow in the ulna. The wrist end of the ulna has a rounded surface that fits into a hollow in the radius. The axis of the hinge that allows us to turn our palms up or down runs obliquely through the elbow end of the radius and the wrist end of the ulna. [see Figure 8] Only one degree of freedom is involved.

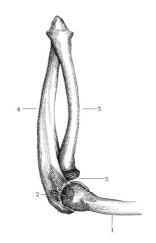

*Figure 8 Medial view of skeleton of forearm. (1) humerus, (2) axis of humerus–ulna joint (3) radius, (4) ulna, (5) axis of radius–ulna joint*

So far we have found joints allowing one and three degrees of freedom. The wrist joint allows two. I can bend and extend my wrist [Plate 42], or I can turn it towards the thumb side or the little finger side. Joints like this allowing two degrees of freedom of movement are called universal joints. The best known examples in engineering are in the transmission systems of motor vehicles. For comfort and safety, a vehicle's wheels must be mounted on a springy suspension that allows them to move up and down relative to the chassis. Universal joints in the drive shaft that connects the engine to the wheels make it possible for power to be transmitted effectively to them, while allowing them to move up and down.

## WRIST AND HAND BONES

The bones of the wrist are known as the carpals. There are eight of them, in two rows of four. [Plate 43 and Figure 9] They are all different shapes but they fit neatly together and are bound to each other by ligaments. Movements of the wrist move each of the carpals relative to its neighbors. Photographs of a skeleton cannot be relied on to show us these movements, because the wires that hold the bones of the skeleton together are not arranged in the same way as the ligaments on the living bones. X-ray pictures of wrists show that turning towards the thumb side or little finger side depends mainly on movement between the bones of the forearm and the first row of carpals. Bending and extension of the wrist involves movement between the forearm bones and the first row of carpals, and between the two rows, in roughly equal measure.

The bones of the palm of the hand are called metacarpals, and the bones of the fingers are phalanges. [Plate 44 and Figure 9] There are five joints in the hand that allow two degrees of freedom each. One of them is at the base of the thumb, where the thumb's metacarpal articulates with the carpal bones. Our ability to grasp things depends critically on this joint. The other universal joints in the hand are the joints between metacarpals and phalanges, where the four fingers meet the palm. You can bend these joints so that the fingers are at right angles to the palm, and you can also spread your fingers out like a fan. The rest of the joints in the fingers and thumb are hinge joints.

Joints are worked by muscles, so some knowledge of muscles helps us to understand bones. Studying the skeleton without reference to the muscles that move it would be like studying the body of a car without considering the engine. The principal muscles of the elbow are in the upper arm, and attach to the radius and ulna close to the elbow. The muscles of the wrist are in the forearm. The palm of the hand might seem the obvious place for the muscles of the fingers, but only a small proportion of them are there. The rest are in the forearm, connected to the finger bones by long tendons. If you tense your hand so as to straighten your fingers as much as possible,

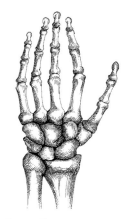

*Figure 9 Hand skeleton.*

*Plate 40 In the palm-down position, the radius crosses over the ulna. (Courtesy Division of Anthropology, American Museum of Natural History.)*

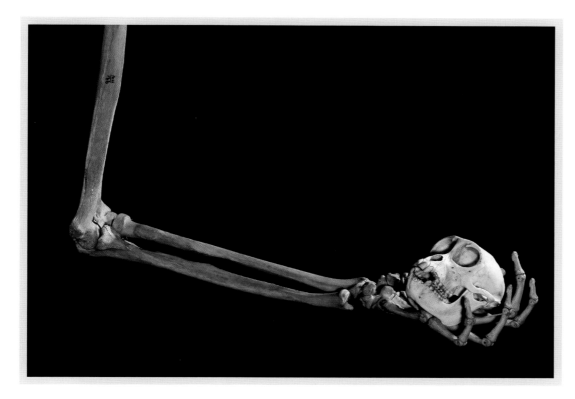

*Plate 41 In the palm-up position, the radius and ulna are parallel to each other. (Courtesy Division of Anthropology, American Museum of Natural History.)*

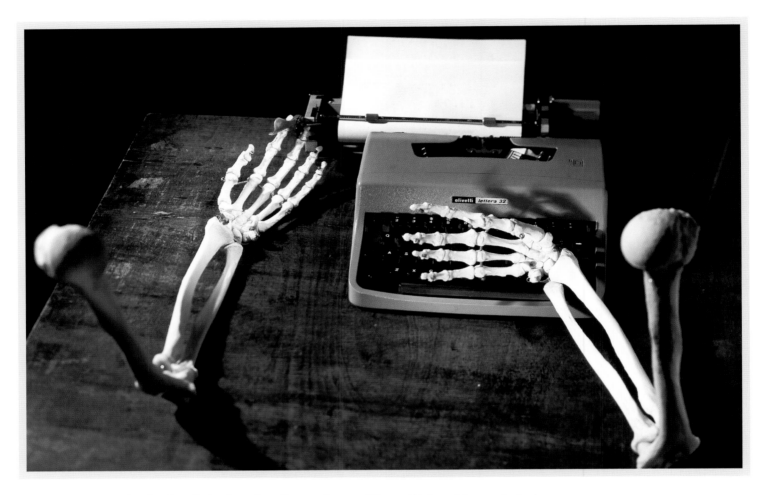

the tendons in the back of the hand will stand out. You will be able to feel them through the skin, with your other hand. An amputated arm that a colleague and I dissected had 20 muscles with a total mass of 91 grams in the hand, and another 20 muscles with a mass of 766 grams in the forearm. Of the muscles in the forearm, nine with a total mass of 401 grams were connected to the fingers and thumb. If all those hand muscles had been in the palm of the hand, the palm would have been swollen to a most awkward shape.

*Plate 42 Wrists bent and extended. (Courtesy Maxilla & Mandible.)*

*Plate 43 (opposite) The skeleton of a hand, showing the bones of the wrist. (Courtesy Maxilla & Mandible.)*

## LEG BONES

Our legs are built on the same basic plan as our arms, which should not surprise us; all four limbs were legs in our ancestors. In the thigh we have the femur, corresponding to the humerus in the upper arm. In the lower leg, the tibia and fibula correspond to the radius and ulna. [Plate 45] In the ankle and foot, the tarsal and metatarsal bones correspond to the carpals and metacarpals, and the five toes to the thumb and fingers. [Plate 46]

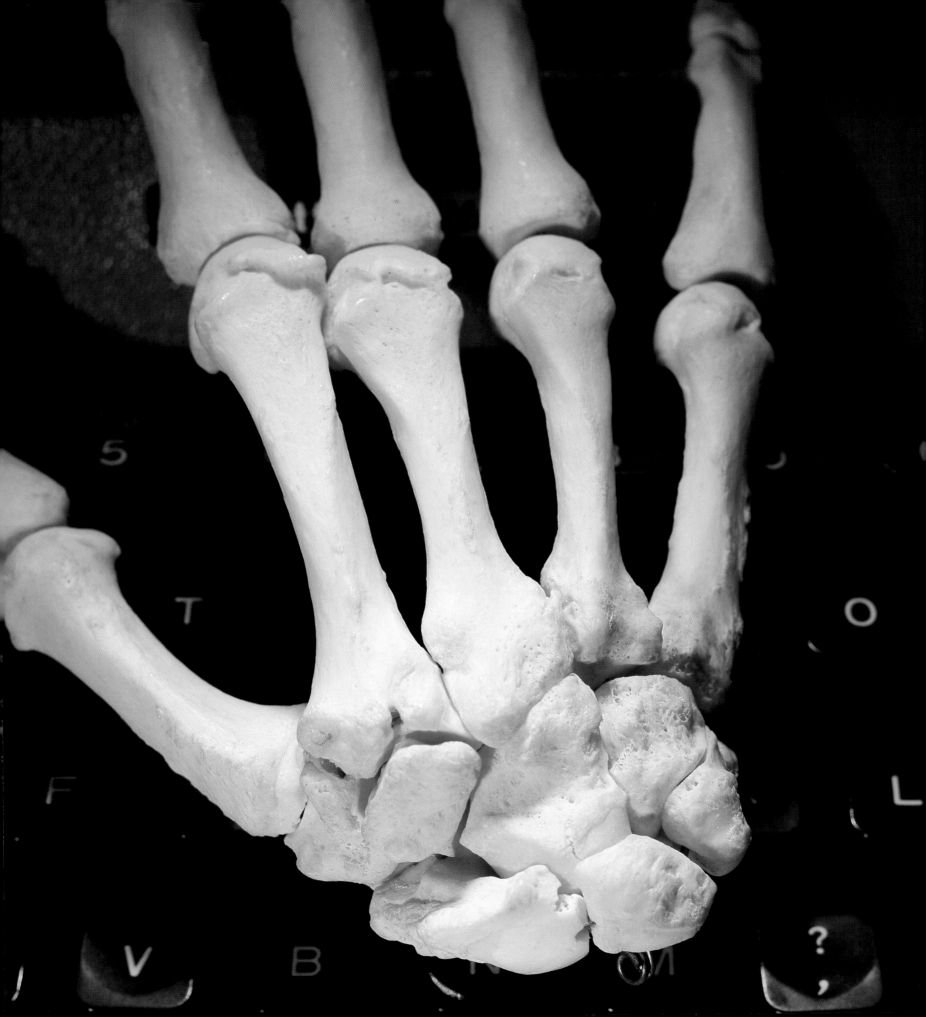

With one exception, the leg joints allow similar movements to the corresponding arm joints. The hip is a ball and socket joint like the shoulder, the knee is a hinge like the elbow and the ankle is a universal joint like the wrist. The exception is in the lower leg. The tibia and fibula cannot cross and uncross in the way the radius and ulna do when we turn our hands palm down or palm up. If they could, we could get our feet into amazing positions. To get an impression of this, sit in a chair with your thighs horizontal and parallel to each other, and your shins vertical. There is enough mobility in the ankle to allow you to rotate your feet, pointing your toes in or out, but only through small angles. If the tibia and fibula could cross and uncross like the radius and ulna, you could rotate your feet through 180 degrees.

Though the leg joints allow broadly similar movements to the corresponding arm joints, there are striking differences in structure. Similarly in quadrupedal mammals, corresponding joints in the fore and hind legs differ. The socket in the pelvic girdle is much deeper than the one in the scapula, making dislocation less likely. [Plate 47] Also, there is a sort of safety rope that the shoulder does not have. This is a ligament connecting the slightly rough area in the middle of the head of the femur to a hollow within the hip socket. This ligament is slack enough to allow the normal range of hip movements, but becomes taut if an excessive upward force on the leg threatens to dislocate the joint.

Though both are hinges, knees (both of quadrupedal mammals and of ourselves) are very different from elbows. In the elbow, the end of the humerus is held in a close-fitting hollow in the ulna. In the knee, however, there is no close fit. [Figure 10 and see Plate 45] Curved surfaces on the end of the femur rest on the almost flat top surface of the tibia. The gaps between the two bones are filled in by wedges of cartilage called menisci, which are held in place by ligaments. There are ligaments connecting the femur to the tibia on the inner face of the knee, and connecting it to the tibia and fibula on the outer face. Also, there are two ligaments that cross like a letter X in a slot in the end of the femur, connecting it to the tibia. These are the cruciate ligaments. As an engineering design, the knee seems clever but rather unsatisfactory. Of all the joints in the body, it is the one most often damaged. Torn menisci and cruciate ligaments are especially common. Queen Elizabeth had a knee operation recently, to remove a fragment of a broken meniscus.

Another difference between knees and elbows is that we have a patella (kneecap) on each knee, but no elbowcaps on our elbows. The kneecap's job is to transmit force from muscles around our knees. The muscles that extend the knee are attached to it, and a ligament attaches it to the tibia. In contrast, the muscles that extend the elbow attach directly to the projection of the ulna behind the elbow, commonly known as the funny bone. (The tingling feeling we sometimes experience when we knock the funny bone is due to a nerve on the surface of the bone being hit.)

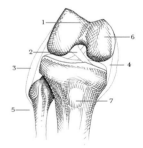
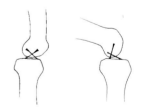

*Figure 10 Front view of bent knee, and diagram showing how the cruciate ligaments make the femur roll back on the tibia as the knee bends. (1) groove for patella, (2) cruciate ligaments, (3) collateral ligaments, (4) meniscus, (5) fibula, (6) femur, (7) tibia*

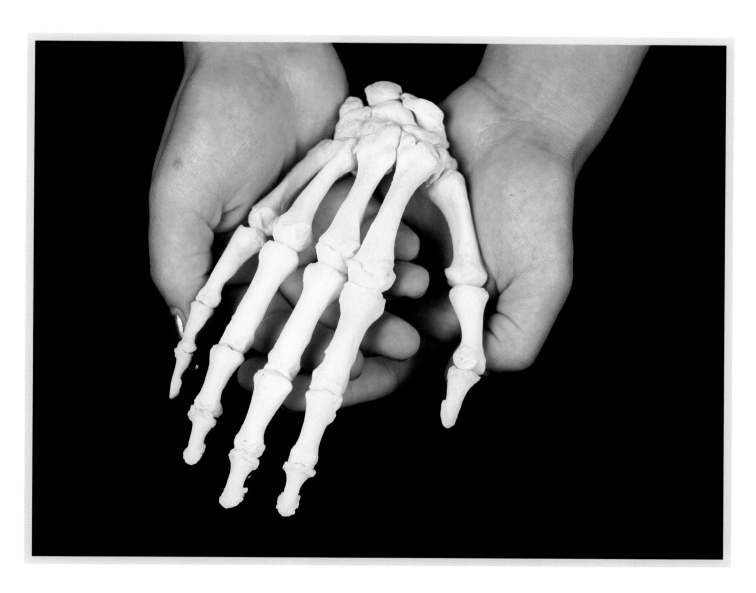

PLATE 44
*The hand skeleton again.*
*(Courtesy Maxilla & Mandible.)*

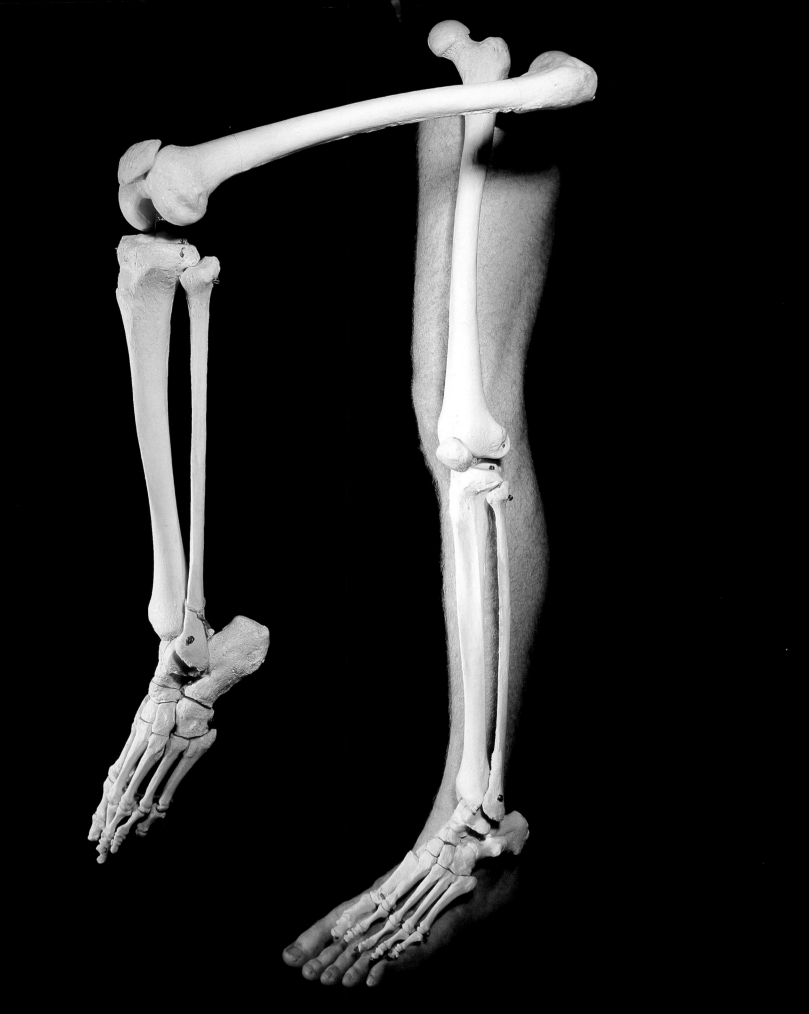

It seems likely that the funny bone evolved from a separate patella-like bone that became firmly fused to the ulna. Birds show us that this suggestion is plausible. Some birds have elbowcaps in their wings, and others have funny bones on their knees. There are no obvious functional reasons for these differences between knees and elbows. It seems that evolution just happened to find different ways of achieving essentially the same results.

## ANKLE AND FOOT BONES

The ankle is a universal joint like the wrist. You can extend and bend it, for example when you stand on tiptoe. You can also rotate it around the long axis of the foot, tilting the foot to one side or the other. The key to these movements is the talus, the only one of the seven tarsal bones that articulates directly with the tibia. [Plate 48 and Figure 11] When we bend and extend our ankles, most of the movement is between the tibia and the talus. When we tilt the foot to one side or the other, most of the movement is between the talus and the other tarsal bones. One of these other tarsals is the calcaneus, which projects at the back of the foot to form the heel. Just as the muscles that extend the elbow attach to the funny bone, the muscles that extend the ankle attach to the calcaneus. They attach to it by means of the Achilles tendon, the big tendon whose elasticity puts a spring into our step, saving energy by making us bounce like rubber balls when we run.

Our feet are arched. Unless you are flat-footed, the prints of your wet feet on the bathroom floor show a gap between the heel and the ball of the foot. Ligaments under the tarsal and metatarsal joints maintain the arch. The large forces that act on the foot in running, however, stretch these ligaments and flatten the arch. When I rest my bare foot lightly on the ground, my ankle is about eight centimeters off the ground. Films show that when the foot is on the ground in running, the foot is squashed down and the ankle gets about a centimeter closer to the ground than that. The stretched ligaments recoil as the foot leaves the ground again, supplementing the spring action of the Achilles tendon.

## REQUIRED JOINTS

We have seen a lot of joints in the arm and leg. How many are actually needed? That depends, of course, on what you want to be able to do. Think first about the leg. You might reasonably want to be able to move your foot in any direction. Movement in any direction can be described as a combination of movements forward or back; to left or right; and up or down. Thus three degrees of freedom of movement are needed. You might also want to be able to rotate your foot in any direction, and we have already seen that rotation in any direction (as at a ball and socket joint) requires three degrees of freedom. In total, we would like to have six degrees of freedom of movement of

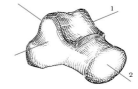

*Figure 11 The talus, showing the axes of its joints. (1) axis of tibia–talus joint, (2) axis of subtalar joint*

*Plate 45 (opposite) The skeleton of the legs. (Courtesy Maxilla & Mandible.)*

the foot relative to the trunk. All these freedoms will inevitably be limited in extent, because human joints cannot rotate full circle, like wheels on axles.

How many degrees of freedom of movement do we have in the leg? There are three in the ball and socket joint at the hip, one in the hinge joint at the knee and two in the universal joint at the ankle, precisely the six that are wanted.

The arm has one more degree of freedom due to the joint between the radius and ulna, giving a total of seven. It may be argued that it has even more than seven, because I have ignored the movements that the scapula can make, relative to the rib cage. Having more than six degrees of freedom may seem unnecessary, but it can be useful. Some of the robot arms that are used for tasks such as assembling cars have only the minimum six degrees of freedom, but some have seven. Suppose you wanted to reach something behind a pillar, but you could not reach it from the right because of some obstruction where your elbow needed to be. You might find it useful to have the option of doing the same job by reaching round the left side of the pillar. Extra degrees of freedom give that kind of flexibility.

The many joints in the hand give it the capacity to do several different things at once. For example, as you tie the ends of a piece of rope together, you will probably find that each hand is simultaneously holding or pushing the rope in three different places.

## JOINT LUBRICATION

In a dry skeleton, all the other tissues have been cleaned off the bone. In the living body, however, the bone surfaces that meet at the joints are coated with slippery layers of cartilage. Cartilage is a jelly formed from water and proteoglycans (which are proteins combined with polysaccharides) reinforced by collagen fibers. Water can be squeezed out of it under pressure, but is drawn back in when the pressure is removed. In an intact joint the cartilage is bathed in synovial fluid, which is mainly water but contains enough protein and polysaccharide to make it viscous like motor oil. A membrane enclosing the whole joint prevents the fluid from leaking away.

This arrangement ensures that the joints are well lubricated. The problems of lubrication in human joints are rather different from the problems of lubrication generally encountered by engineers. In machines, shafts usually revolve steadily in one direction. They drag the viscous oil round with them in their bearings, so that oil is constantly being drawn under the shaft and the metal of the shaft never rests directly on the metal of the bearing. This is the principle of hydrodynamic lubrication. Human joints cannot make even a single complete revolution, but move

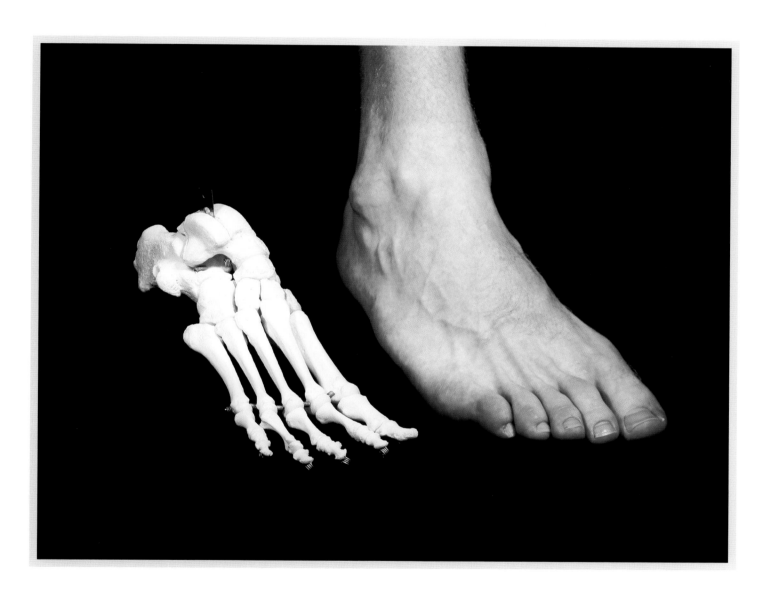

PLATE 46
*The skeleton of a foot.*
*(Courtesy Maxilla & Mandible.)*

backwards and forwards through relatively small angles. Hydrodynamic lubrication cannot work well for them.

Despite a lot of research effort, the mechanism of lubrication of human joints is still not entirely clear. In many activities, the forces on joints are intermittent. For example, large forces act across the joints of a runner's leg while the foot is on the ground, but the forces are small while the leg is being swung forward for the next step. Some of the synovial fluid gets squeezed out from between the bones while the foot is on the ground, but the interruption of the force gives the fluid the chance to seep back in. The cartilage surfaces are not perfectly smooth, but have microscopic dimples that trap the fluid, making it harder for the fluid to be squeezed out. Because cartilage is elastic, cartilage surfaces that are pressed together deform, increasing the area of near-contact. This also makes it harder for the fluid to be squeezed out. Water from the fluid between the cartilage surfaces may get squeezed out through pores in the cartilage, but the protein and polysaccharide molecules in the fluid are too big to get through the pores, so squeezing out makes the lubricant more concentrated. In the knee, the rounded surfaces of the femur roll backwards and forward over the flatter surface of the tibia as the joint bends and extends, so while movement continues no part of the cartilage remains under load for long.

Whatever the explanation (or explanations), joints are very effectively lubricated. Most experiments have been done on joints removed from dead animals, but a clever technique makes it possible to investigate the lubrication of a joint in living people. Lay the back of your hand flat on a table. Now, keeping all the other joints straight, bend the middle joint of your middle finger through a right angle, so that the end of your finger points vertically. If the muscles and tendons of your hand are arranged in the usual way, you will find that you have lost control of the terminal joint of the finger. Touch the fingertip with the other hand and you will find that it wobbles freely. In a series of experiments, a subject's hand was strapped down in the position I have described and a pendulum was suspended from the fingertip. The pendulum was set swinging, and its movements were recorded while its oscillations died down. From the observations, the coefficient of friction of the joint was calculated; a low coefficient of friction indicates good lubrication. Coefficients of friction measured in this experiment and in others with animal joints were of the order of 0.01, about the same as for good engineering joints.

The human skeleton is a product of evolution by natural selection. We have seen in this chapter that many of the solutions to mechanical problems, that natural selection has found, have parallels in engineering. The tubular structure of limb bones and scaffolding poles gives both the strength and stiffness that they need, without making them unnecessarily heavy. Bone and fiberglass are both composite materials, benefiting from a principle that gives them mechanical

*Plate 47 (opposite) The head of the femur fits a deep socket in the pelvic bone. (Courtesy Maxilla & Mandible.)*

85

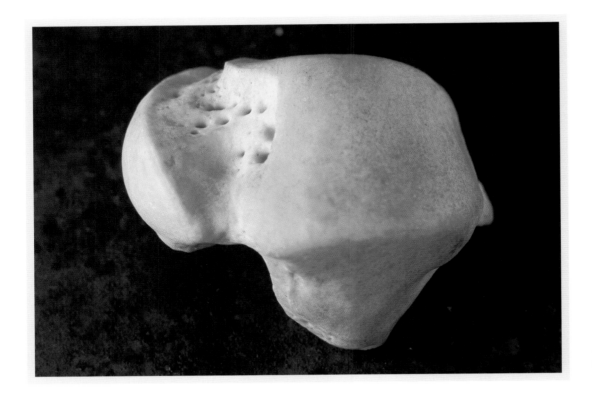

Plate 48 *The talus seen from above. The toes would be off the picture to the left. (Courtesy Maxilla & Mandible.)*

Plate 49 *The man-made arm of the digger is jointed like a human arm. (Courtesy Maxilla & Mandible.)*

properties far superior to the ingredients that are blended to make them. Skeletons and machines have hinge joints, universal joints and ball and socket joints, which are combined in similar ways in human arms and the arms of robots and mechanical diggers. [Plate 49]

It is tempting to present the skeleton as a triumph of engineering design, perfected over millions of years by evolution. That would be naïve. The network of spongy bone in the head of the femur (see Plate 37) is beautifully aligned with the stresses that act in it when we walk or run, but that is not the direct result of natural selection. Evolution works only on features that are programmed in our genes. The details of the spongy bone's design are not dictated by our genes, but develop in response to the stresses that act on it. The processes of bone growth and replacement, which we saw in action in Chapter 1, automatically adapt themselves to patterns of stress.

In many respects, our limb skeletons seem beautifully adapted to their functions, but many details seem impossible to explain by function alone. As in the skull, we must consider ancestry as well as function. For example, our forearms and lower legs have two bones each: the radius and ulna, and the tibia and fibula. There is no apparent need for two bones in either case, but those bones can be traced back through 400 million years of evolution. They were already present in the fins of our fishy ancestors. As another example, I could not honestly argue that five is the ideal number either for fingers or for toes. Fore and hind legs evolved from fins 350 million years ago, in the earliest amphibians. Five was soon established as the normal number of fingers and toes. Some animals have since lost toes in the course of evolution (for example, cows have only two toes left on each foot, and horses only one), but we have retained all five. We have not retained five toes, and two lower leg bones, because those are necessarily the best numbers, but because those are the numbers our ancestors had. Evolution often leaves things as they are when there is no strong reason for change.

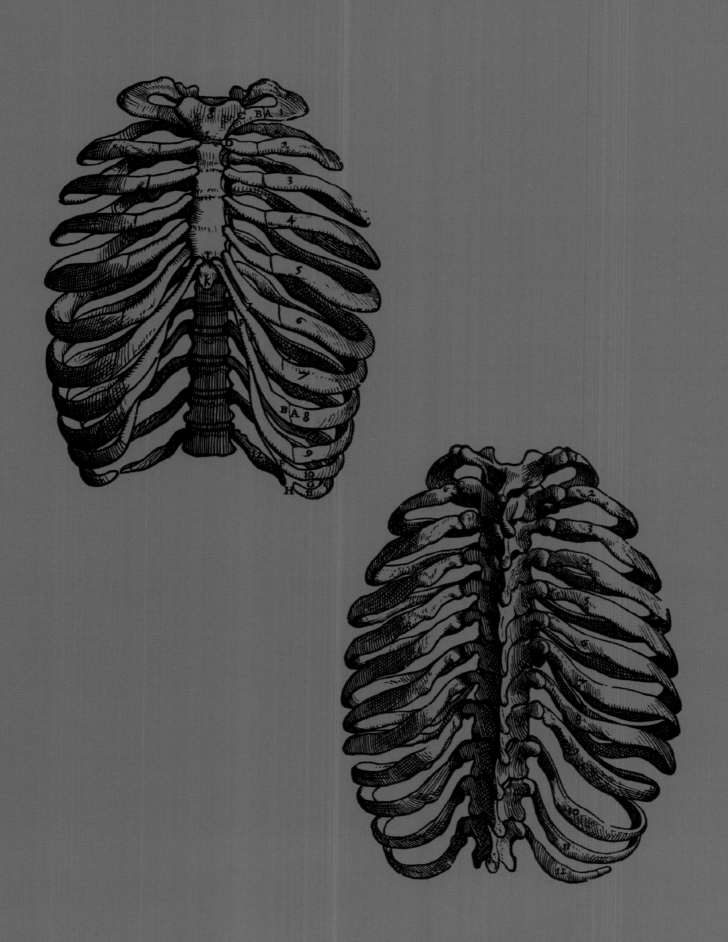

# 4

# THE TORSO

The torso is the core of the body, everything except the arms, legs and head. [Plate 50] It is built around the backbone: seven cervical vertebrae in the neck, twelve thoracic vertebrae in the upper back, five lumbar vertebrae in the lower back and finally the sacrum and coccyx, each of which consists of several vertebrae fused together into a single block of bone. [Plate 51, which omits the sacrum and coccyx] I have found it convenient to include the shoulder girdle and pelvis in this chapter, as parts of the torso, though they might equally well be regarded as parts of the limbs.

## VERTEBRAE

Each vertebra has the same basic structure. [Plate 52 and Figure 12] There is a cylindrical block of bone, the centrum, which is the main load-bearing component. Behind it (closer to the skin of your back) is the neural arch. The neural arches of successive vertebrae form a protective tunnel for the spinal cord. Gaps between the neural arches serve as passageways for the spinal nerves. The apex of the neural arch carries a narrow blade of bone, the neural spine, which separates the muscles of the left side of the back from those of the right.

Two short arms of bone, the transverse processes, stick out on either side of the neural arch. They lie between the muscles that bend the back and those that straighten it. The psoas muscles, which bend the back, lie on either side of the centra in front of the transverse processes. Fillet of beef is the psoas muscle. The much larger erector spinae muscle, which straightens the back, lies behind the transverse process. Sirloin of beef

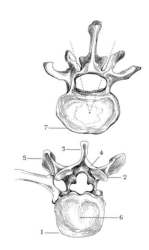

Figure 12 *Two vertebrae. (1) Thoracic vertebra, (2) zygapophysis, (3)neural spine, (4) neural arch, (5) transverse process, (6) centrum, (7) Lumbar vertebra*

is part of the erector spinae muscle. The bone in a T-bone steak is the transverse process of a lumbar vertebra. In pork and lamb chops, the bone is half a vertebra and a rib, and the large disc of meat is a slice of the erector spinae.

The joints between successive vertebrae allow us to bend and twist our backs. They are quite different from the joints in our arms and legs. In limb joints, the cartilage coating on the end of one bone is entirely separate from the cartilage on the other. When we bend and extend the joint, the cartilage of one bone slides over the surface of the cartilage on the other. In contrast, the centra of adjacent vertebrae are connected by an intervertebral disc, which is attached firmly to both. Bending and twisting of the back involves distortion of the intervertebral discs. These consist of proteoglycan (protein plus polysaccharide) jelly enclosed by a thick wall of collagen fibers, so are reasonably flexible. X-ray pictures of normal, healthy people show that when we bend our backs as much as possible, each lumbar vertebra is bent at an angle of 12-15 degrees to the next. Small angles between each vertebra and the next add up to allow the back as a whole to bend through a large angle. Contortionists can bend through unusually large angles. Richard Wiseman of the University of Hertfordshire has found a possible explanation. He took magnetic resonance pictures of a contortionist's spine and found that her intervertebral discs were unusually thick.

Bending and twisting of the joints between the vertebrae is restricted by flanges of bone called zygapophyses. Zygapophyses on the lower edges of the neural spine of one vertebra rub against zygapophyses that project up from the next vertebra below. The surfaces that rub together are covered with cartilage, and form low-friction joints. The centra have just a thin skin of compact bone. Inside they are spongy bone, like the bone inside the ends of limb bones. Spongy bone is very much weaker than compact bone, weaker even than you might guess from its composition. Tests on samples of bone from cows and people have shown that spongy bone that is half solid bone and half empty space is only about one quarter as strong as compact bone. If the fraction of solid bone is reduced to one quarter, the strength goes down to about one sixteenth.

This suggests that our vertebral centra could have been made much more slender, from less bone, if they had been made compact instead of spongy. A slender backbone, however, would be in danger of buckling under load. Try building a tower from a child's building blocks, using just one block for each storey. The bigger the blocks, the taller the tower that you are likely to be able to build. That analogy is not entirely fair because vertebrae, unlike building blocks, are connected by intervertebral discs, but even solid pillars are liable to buckle if they are too slender.

## VERTEBRAE STRENGTH

Typical men weigh about 70 kg and typical women about 50 kg. The centre of gravity of the body, in both sexes, is a few centimeters above the hip joints. That means that

*Plate 50 (opposite) The skeleton of the torso. (Courtesy Maxilla & Mandible.)*

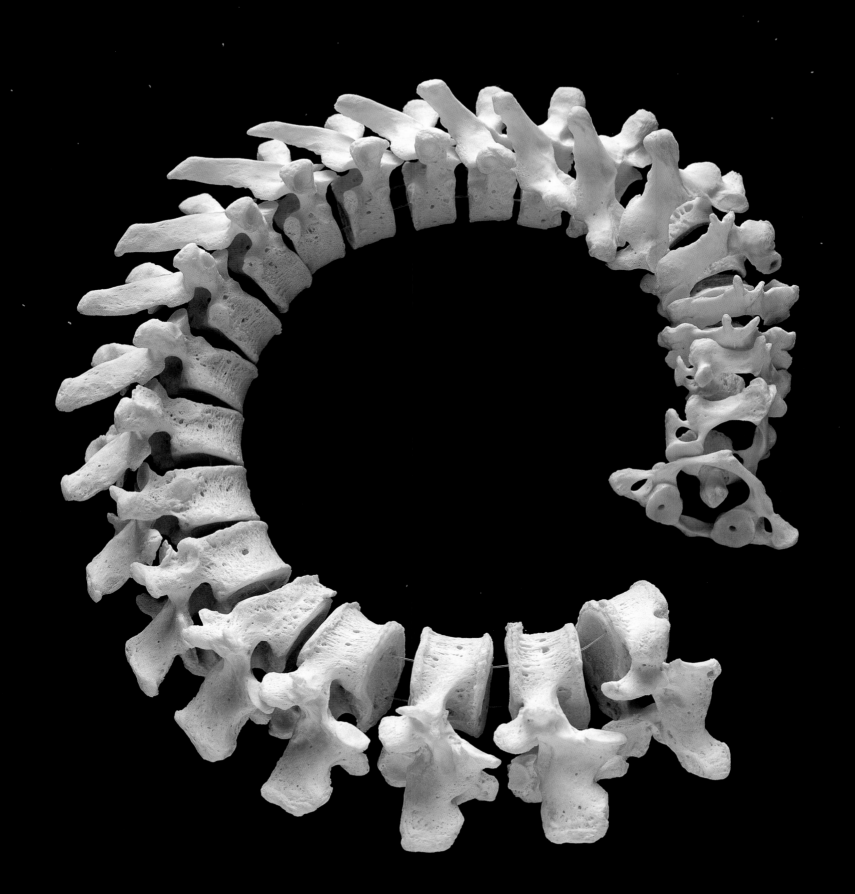

the lower lumbar vertebrae have to support about half the weight of the body, 35 kg in a man and 25 kg in a woman. Those loads act when we sit or stand quietly, but in other circumstances the forces on the vertebrae are very much larger. The best middleweight male weightlifters (67.5-75 kg) can lift about 200 kg in a clean and jerk. When they are supporting that weight, the load on the lower lumbar vertebrae (the weight of the barbell plus half body weight) is about 235 kg. Considerably larger loads act at some stages of the lift.

Only exceptional athletes can lift such heavy weights, but huge loads act on the vertebrae of ordinary people when their back muscles are active. One set of experiments used a force transducer (an instrument for measuring forces) bolted to the floor. This transducer had a handle attached to its top. Healthy young men were asked to bend over and pull upwards on the handle. When they pulled as hard as they could, the transducer registered forces of (on average) 90 kg. Think about the balance of forces in the lower back. The line of action of the force on the transducer was well in front of the spine, but the balancing force (in the erector spinae muscles) was close

*Plate 51 (opposite) The vertebral column, omitting the sacrum and coccyx. (Courtesy Maxilla & Mandible.)*

*Plate 52 (above) A lumbar vertebra, seen from above. (Courtesy Maxilla & Mandible.)*

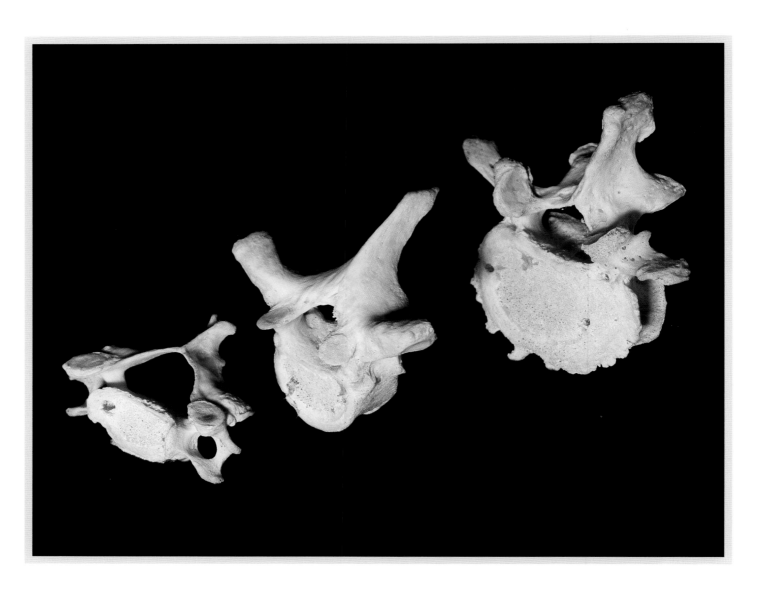

PLATE 53
*Typical cervical, thoracic and lumbar vertebrae*
*(from left to right.)*
*(Courtesy Maxilla & Mandible.)*

behind the spine. By the principle of levers, the muscle force was many times greater than the force on the handle. The force in the muscles was calculated to be about 740 kg, and the total load on the lumbar vertebrae 800 kg, or 0.8 tons. The accuracy of calculations like these has been checked by experiments in which the pressures inside the intervertebral discs of volunteers have been measured through hypodermic needles.

In other experiments, the strengths of lumbar vertebrae from human cadavers have been measured by crushing them in engineers' testing machines. Many of them were found to be far too weak to withstand the forces calculated from the lifting tests. For obvious reasons, most of the vertebrae came from elderly people who had been relatively inactive for some time before they died, so they may have been weaker than they were when the people were young and healthy. One set of experiments used only vertebrae from men aged 46 or younger, who had been mobile immediately before death. The average strength of these vertebrae was about 1000 kg (one ton).

Good sense and safety regulations require engineers to design structures to be considerably stronger than the greatest forces expected to act on them. The safety factor is the design strength divided by the expected maximum load. Bridges have commonly been given safety factors of about two, and the cables of passenger elevators about ten. Substantial safety factors are advisable for two reasons. Variations in workmanship or in the quality of the materials may result in a structure being weaker than it was intended to be. And circumstances may arise that were not anticipated, in which the loads are greater than expected. If the lumbar vertebrae of healthy young men were capable of withstanding only 1000 kg, and had to withstand 800 kg in heavy lifting, their safety factor would be 1.25, a dangerously low value. Much higher safety factors of 2–4 have been calculated for leg bones of animals, as we shall see in Chapter 5. Perhaps even the vertebrae from the younger subjects had been weakened by inactivity during illness before death.

## VERTEBRAE GROUPS

Every vertebra in the back is different from all the others. An experienced anatomist, given a single vertebra, can recognize its exact position in the backbone. The differences between the major groups of vertebrae (cervical, thoracic and lumbar) are obvious [Plate 53], but within each group the differences between immediately adjacent vertebrae are generally quite small. The first two vertebrae, at the top of the neck, are exceptional in being markedly different from the rest. The remaining five are all similar to each other, though the transverse processes get progressively longer towards the base of the neck. The twelve thoracic vertebrae are all pretty much alike, but they increase gradually in size towards the lumbar region, and the angle of the neural spine changes. [Plate 54] Similarly, the five lumbar vertebrae are only a little different from each other.

*Plate 54 (following pages) Each vertebra is just a little different from the next. (Courtesy Maxilla & Mandible.)*

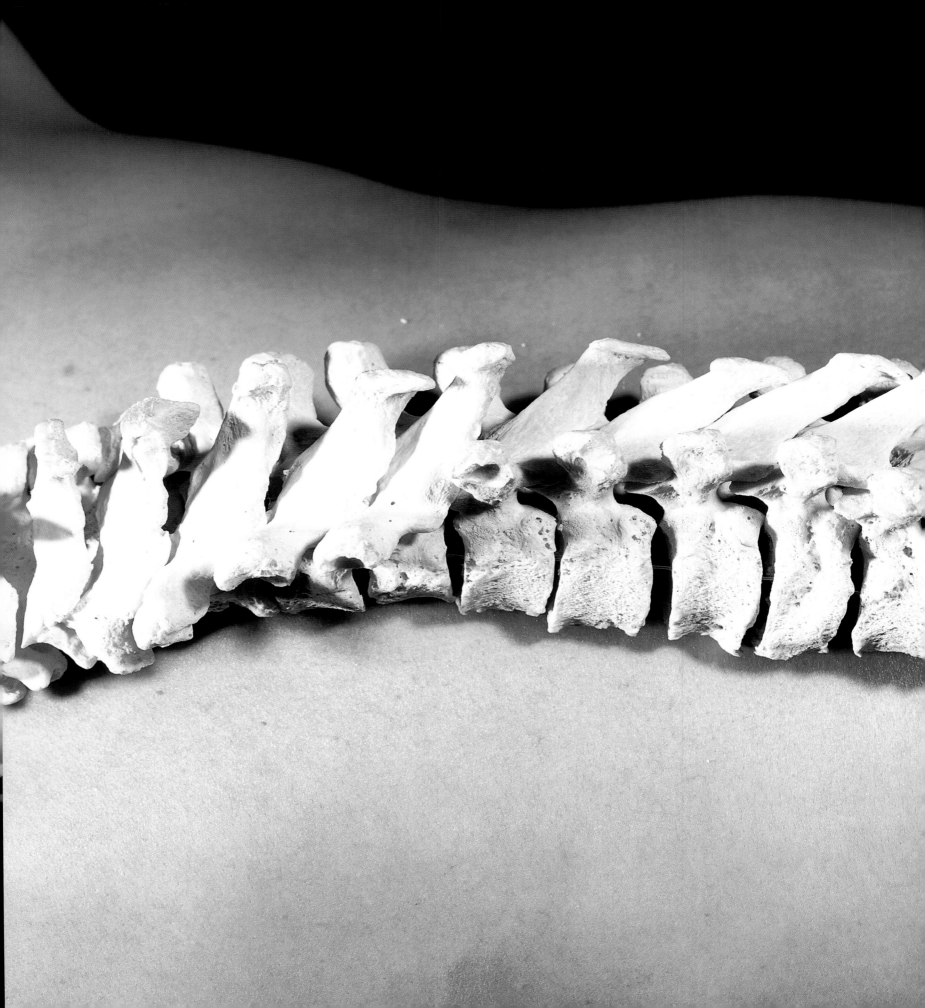

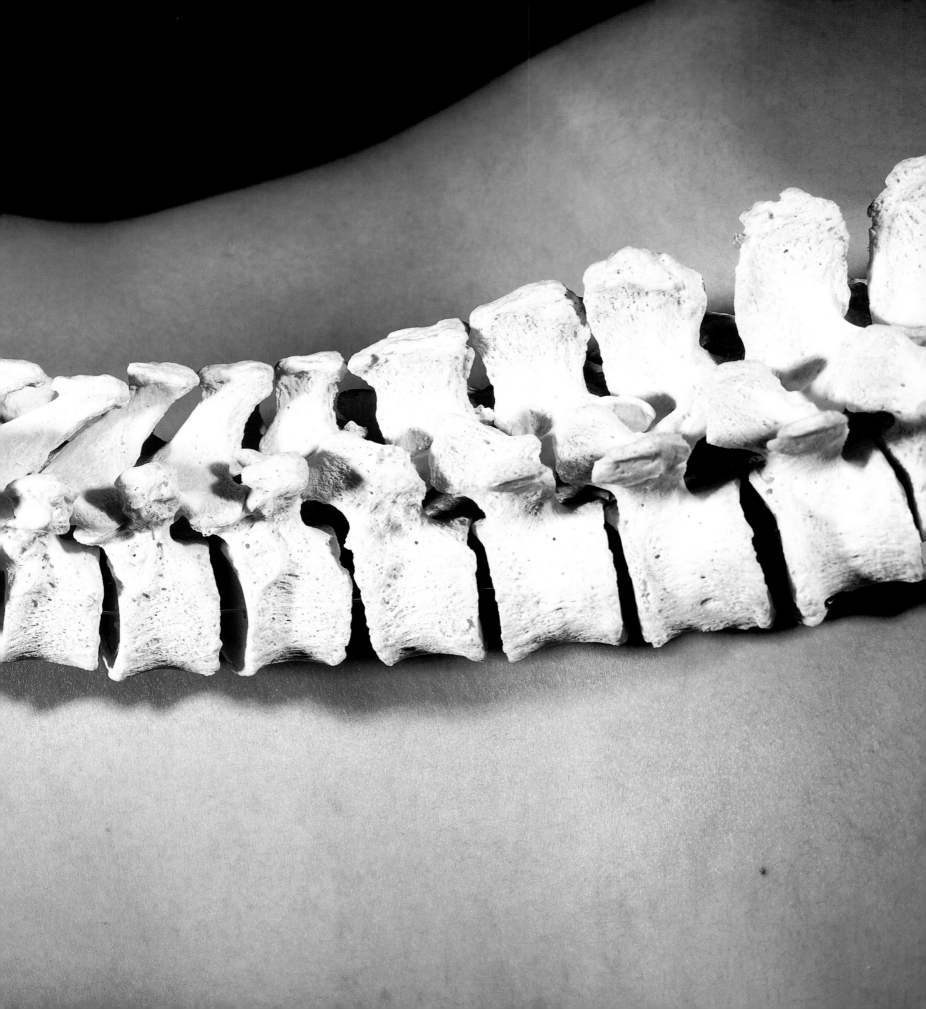

## CERVICAL VERTEBRAE

We will look at the vertebrae in turn, starting in the neck. The first vertebra [Plate 55] is known as the atlas because it supports the head [Plate 56], much as the giant Atlas of Greek mythology held up the heavens. It is very different from the other vertebrae, ring shaped with no centrum or neural spine. It has shallow depressions on either side of its upper surface, to the left and right of the space for the spinal cord. Matching bulges on the underside of the skull rest in these depressions. The bulges and depressions are covered with thin layers of cartilage, forming well- lubricated synovial joints like the joints in the limbs. This arrangement functions as a hinge, allowing the skull to rock on the top of the spine as it does when we nod our heads. The second vertebra, the axis, is quite different in shape. Unlike the atlas, it has both a centrum and a neural spine. An extension of the centrum called the dens reaches forward into the ring- shaped atlas, in front of the spinal cord. A ligament across the ring prevents the dens from pressing against the spinal cord. The zygapophyses are

*Plate 55 (above) The atlas and axis vertebrae, seen from above. (Courtesy Maxilla & Mandible.)*

*Plate 56 (opposite) The skull is supported by the cervical vertebrae. (Courtesy Maxilla & Mandible.)*

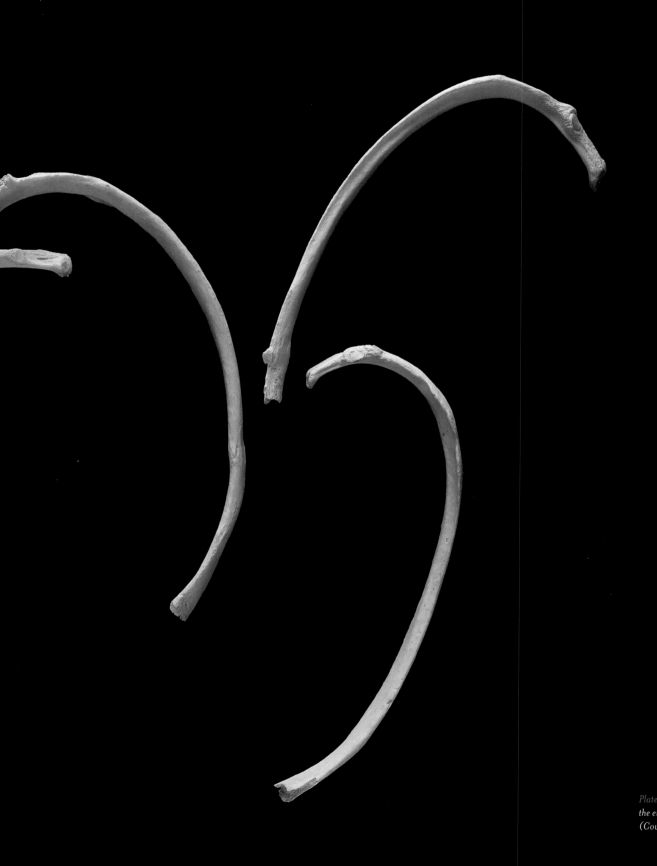

*Plate 57 The ribs curve most sharply at the end that connects to the spine. (Courtesy Maxilla & Mandible.)*

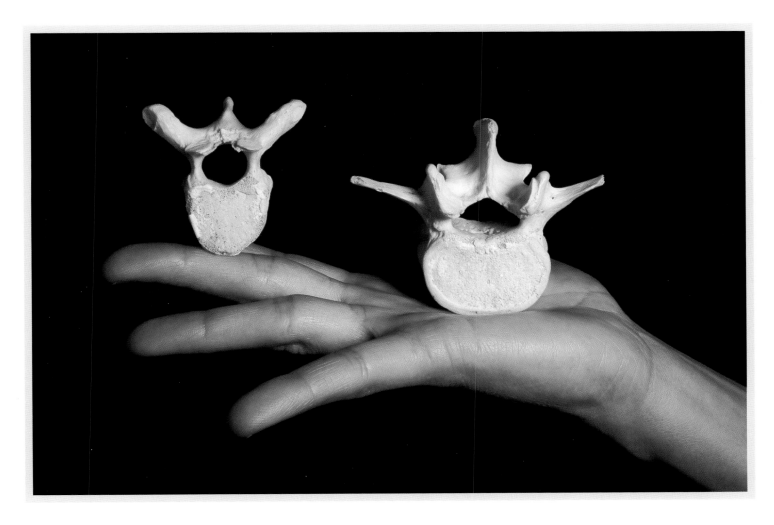

arranged so as to allow the atlas to rotate around the dens. Whereas the joint between the atlas and the skull allows nodding, the one between the axis and the atlas allows us to shake our heads.

The zygapophyses of the remaining cervical vertebrae are set at angles that allow some rotation and a little nodding and side to side bending, between each vertebra and the next. One striking feature of all the cervical vertebrae is that they have projections with holes in them, sticking out on either side of the centrum. These projections are the rudiments of ribs that our reptile ancestors had in their necks.

## THORACIC VERTEBRAE

The characteristic feature of the thoracic vertebrae is that the ribs attach to them. Most of the ribs attach in two places. The head (extreme end) of the rib articulates with a facet on the centrum of a thoracic vertebra. A lump a short distance from

*Plate 58 (opposite) Two successive thoracic vertebrae showing the facets (zygapophyses), which would interlock if the vertebrae were fitted together. (Courtesy Maxilla & Mandible.)*

*Plate 59 (above) A thoracic vertebra (left) and a lumbar vertebra (right), seen from above. (Courtesy Maxilla & Mandible.)*

the end can be seen in Plate 57. It articulates with a facet on the transverse process of the same vertebra. Both these attachments are synovial joints. Because there are two articulations, the joint between the rib and the vertebra functions as a hinge, allowing rotation about an axis that passes through both articulations. The ribs rotate about these axes when we expand and contact our chests, as we breathe. The holes that we saw in the rib-rudiments on the cervical vertebrae correspond to the gap between the two attachments of thoracic ribs.

If the vertebrae were connected only by the intervertebral discs, their joints could bend a little in any direction, and twist a little. The movements might be small, but there would be three degrees of freedom of movement, as in a ball and socket joint. An end view of a thoracic vertebra [Plates 58 and 59 and see Figure 12] shows that the articular surfaces of the zygapophyses lie more or less on the surface of a sphere that has its centre on the axis of the centrum. Each zygapophysis is in effect a tiny part of a ball and socket joint.

Accordingly, the zygapophyses of the thoracic vertebrae do very little to limit movement. Each vertebra can twist through a range of about eight degrees (four degrees to left or right) relative to the next. Eight degrees may not seem much, but eight degrees rotation around each of eleven joints gives a total of 88 degrees. You can check that your back is capable of twisting about that much, by turning your shoulders to the left and right while sitting on a chair. Each thoracic vertebra can bend forward and back relative to the next through a range of about five degrees, and it can bend to left and right through a range of about eight degrees. The smaller of these ranges, for bending forward and back, seems to be limited more by the attachments of the vertebrae to the ribs, than by the zygapophyses.

## LUMBAR VERTEBRAE

The lumbar vertebrae in the lower back have no ribs. Their neural spines are broader than those of the thoracic vertebrae and their transverse processes are longer. [Plate 60] Also, their zygapophyses are set at quite different angles. Instead of lying on an imaginary sphere centered on the centrum, their articular surfaces stand at right angles to the surface of the sphere, radiating out from the long axis of the spine. [see Figure 12]. Arranged like that, they allow very little twisting about the long axis, but they allow the lumbar region to bend freely forward and back, and quite a lot from side to side.

The ranges of movement between successive vertebrae are around two degrees of twisting, 14 degrees of forward-and-back bending and six degrees of side-to-side bending. When you twist your body, nearly all the movement is between thoracic vertebrae, but when you bend to touch your toes most of the movement is between lumbar vertebrae.

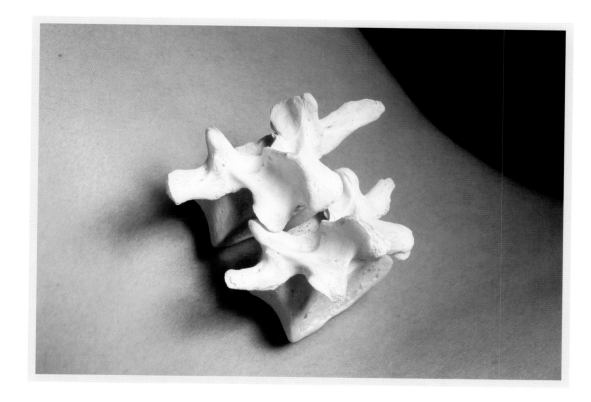

*Plate 60 Two lumbar vertebrae.*
*(Courtesy Maxilla & Mandible.)*

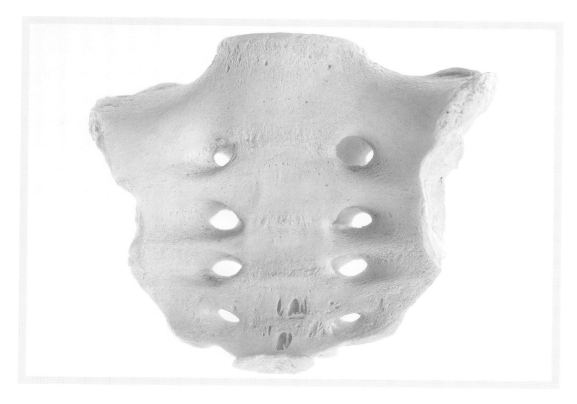

*Plate 61 The sacrum.*
*(Courtesy Maxilla & Mandible.)*

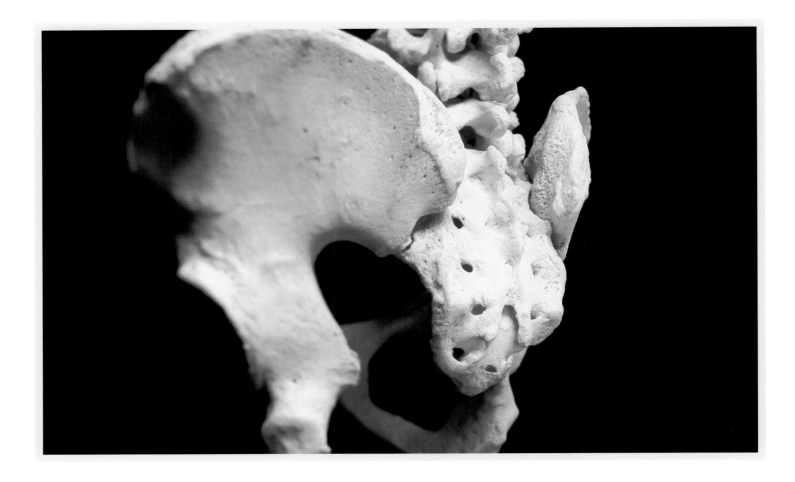

## SACRUM AND COCCYX

Next after the lumbar vertebrae is the sacrum, which consists of five separate vertebrae in children. These vertebrae coalesce after puberty, so that the adult sacrum is a single block of bone. [Plate 61] Even in adults, spaces between the bases of the transverse processes show where the boundaries of the original vertebrae were. Embryological studies show that what appear to be merely transverse processes are actually transverse processes with very short, thick ribs fused to their ends. In reptiles the sacral ribs and vertebrae are separate bones, but in mammals they are fused together. The ends of the sacral ribs are attached to the pelvic girdle. [Plate 62] The last element in the backbone is the coccyx, which is the rudiment of the tail of our monkey ancestors, [Plate 63] It consists of three to five rudimentary vertebrae, far fewer than the 20 or more of a typical monkey's tail. They are usually fused together, but the first one is sometimes separate. Tails are useful in monkeys for balance, and in South American monkeys as an additional grasping limb, but the human coccyx seems to be entirely useless. You may wonder why it has not disappeared in the course of evolution. The answer, probably, is that it does us no harm and costs very little. Growing and maintaining a coccyx uses very little

*Plate 62 (above) The sacrum in place, connecting the pelvic girdle to the backbone. (Courtesy Maxilla & Mandible.)*

*Plate 63 (opposite) The coccyx is in the middle of the picture, at the lower end of the backbone. (Courtesy Maxilla & Mandible.)*

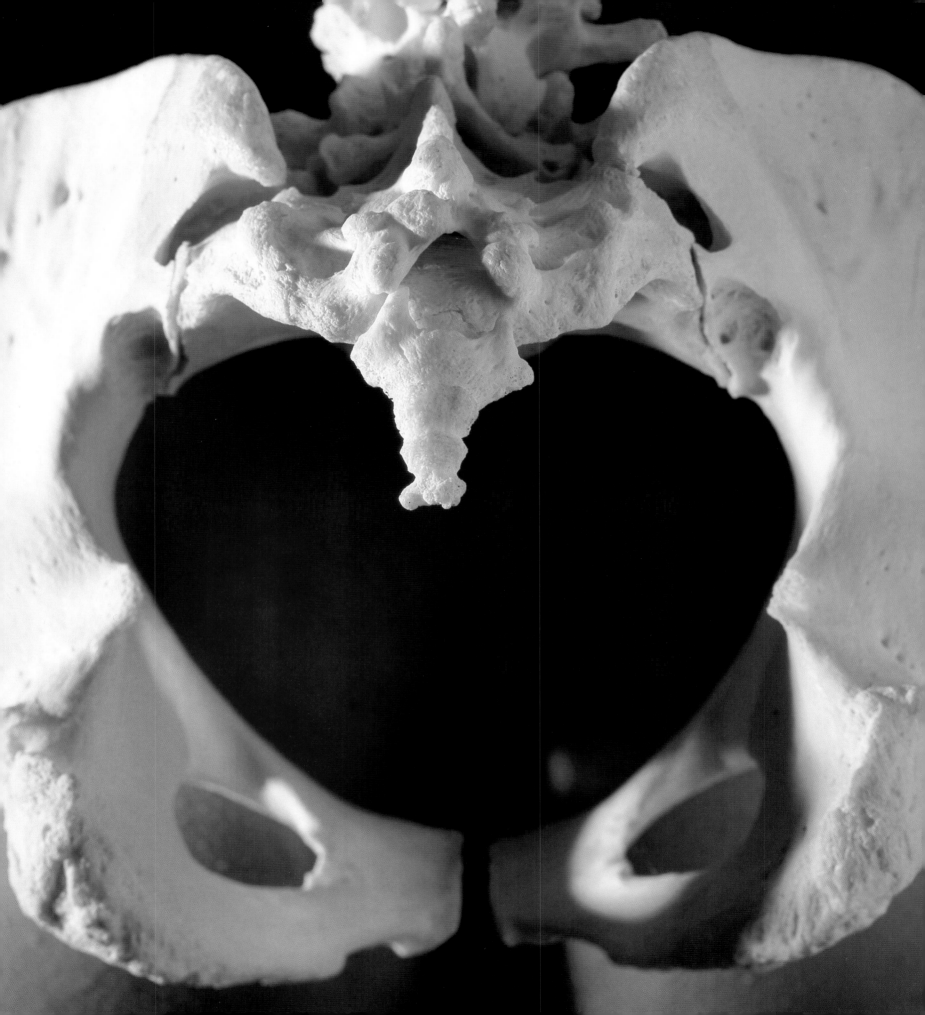

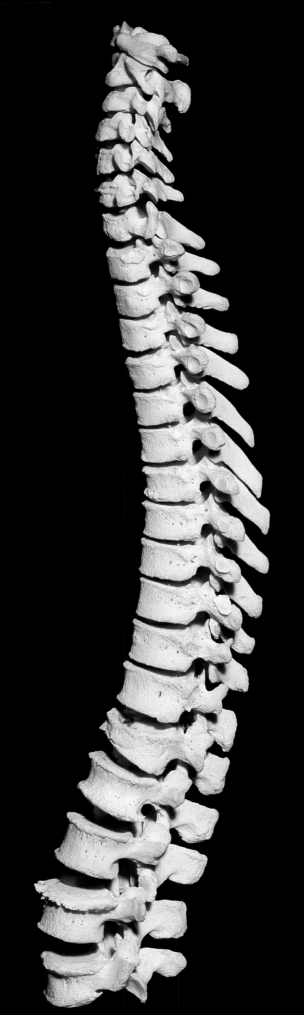

Plate 64. *The vertebral column forms a sinuous curve. The sacrum and coccyx, which form one more (concave forward) bend are missing from this specimen. (Courtesy Maxilla & Mandible.)*

energy or materials, so the advantage of losing it would be very slight. Natural selection for really trivial improvements may be too weak to be effective. The spine as a whole has a sinuous curve. The cervical region is convex forward, the thoracic part is concave forward, the lumbar region convex forward (making the hollow of the back), and the sacrum concave forward again. [Plate 64] The backs of well-designed chairs are curved accordingly, concave against the upper (thoracic) part of a sitter's back and convex against the lower (lumbar) part. The hollow in the back is a specifically human characteristic, not found in other mammals. Its evolution was one of the changes involved in giving us our upright posture.

## RIBS

The ribs form an oval cage, wider from side to side than from front to back, enclosing the heart and lungs [Plate 65, and see Plate 57] They are bone for most of their length, but a section at the front of the chest is cartilage. At the back, they are attached to the thoracic

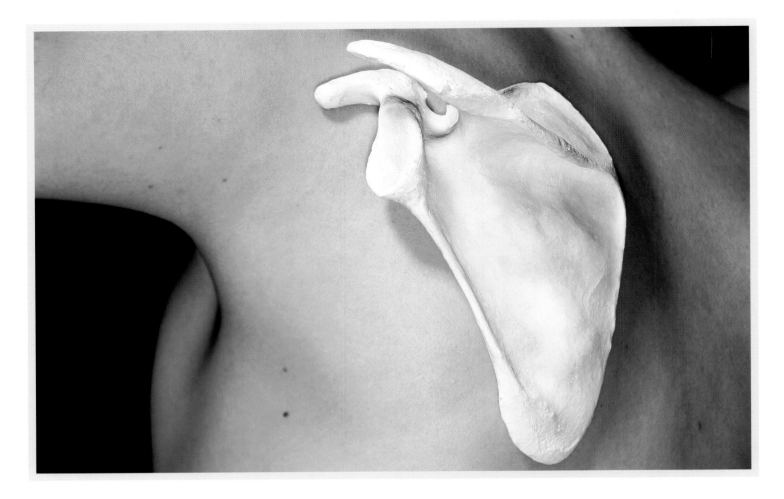

vertebrae, as we have already seen. At the front the cartilaginous ends of most of them are attached to the sternum or breastbone, which is actually a row of three plates of bone.

We expand our chests by moving our ribs. If you put your hands on your chest while you breathe strongly in, you will be able to feel that the ribs rise as the chest expands. There are muscles running from the cervical vertebrae to the first two ribs. [Figure 13] When they contract, they lift the ribs. Also, the gap between each rib and the next is filled with muscle. (The equivalent muscles in pigs are the meat on spare ribs.) The fibers of these muscles run obliquely from one rib to the next. Some of them are angled in such a way as raise the ribs and expand the chest when they contract. Others, sloping the other way, have the opposite effect, pulling the ribs down and reducing the volume of the chest. They are used for breathing out.

Breathing depends only partly on rib movements. The diaphragm is a sheet of muscle that forms a partition across the body, separating the thoracic cavity, which contains the lungs and heart, from the abdominal cavity, which contains the stomach, intestines and liver. [see Figure 13] It attaches to the lower edges of the rib cage, so that the lungs are

*Plate 66 (above) The scapula.*
*(Courtesy Maxilla & Mandible.)*

*Plate 67 (opposite) The pelvis.*
*(Courtesy Maxilla & Mandible.)*

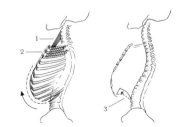

*Figure 13 How rib movements (left) and contraction of the diaphragm (right) enlarge the thorax. (1) scalene muscles, (2) external intercostal muscles, (3) diaphragm*

110

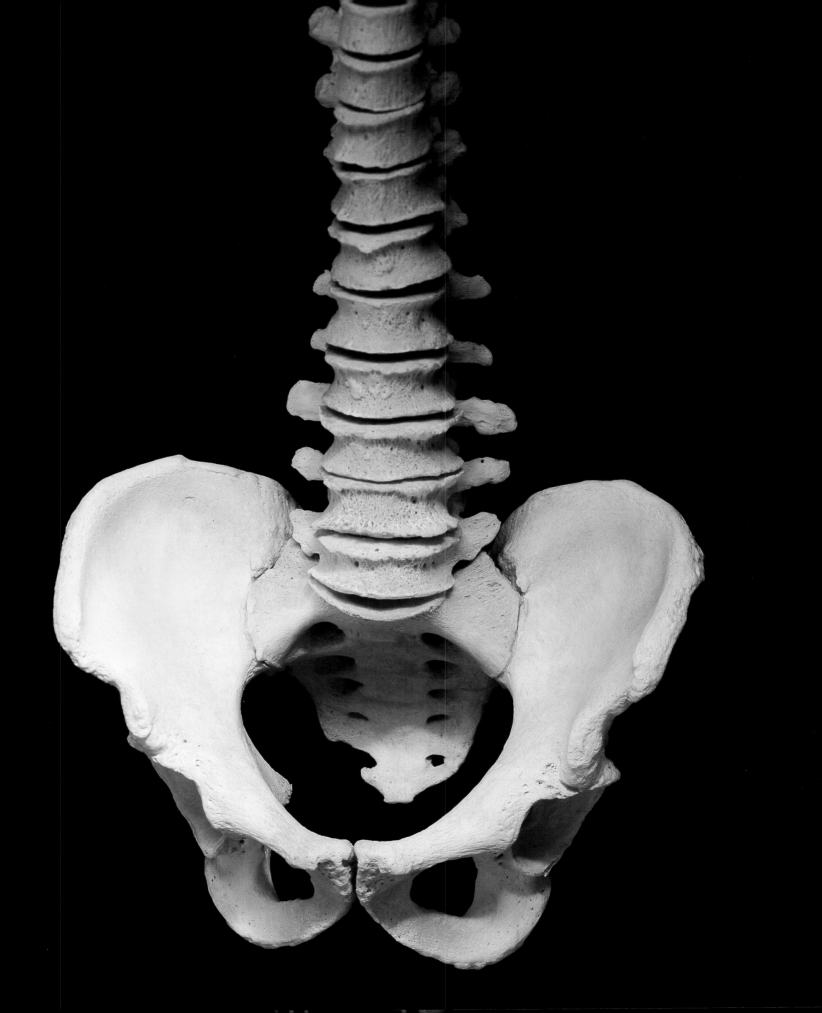

enclosed in a chamber that has rather rigid walls and a muscular floor. The way in which it works can be seen in X-ray pictures. Bones show up in radiographs because bone is less transparent to X-rays than flesh. With careful adjustment of the exposure, the air in the lungs can also be made to show up because air is more transparent to X-rays than flesh. Radiographs show that when we breathe in our lungs expand downwards towards the abdomen. This is due to contraction of the diaphragm, which is strongly domed at rest but flattens as it contacts.

The rib cage is essential to the working of the diaphragm, because it prevents the wall of the chest from collapsing inwards when the diaphragm contracts. In the abdomen, however, ribs would be unhelpful. When the diaphragm contracts and flattens, it displaces the liver and intestines, making the belly bulge out. If the wall of the abdomen were stiffened by ribs, this could not happen so easily. Reptiles, which do not have diaphragms, have ribs all along the length of the trunk, but when the ancestors of the mammals evolved diaphragms they lost their abdominal ribs.

## SHOULDER BLADE

The scapula (shoulder blade) is a thin plate of bone with a keel-like projection (the scapular spine) on its outer face. [Plate 66] It lies flat against the back of the rib cage, attached to the ribs only by muscles. Apart from the humerus, the only bone the scapula connects with is the clavicle (collarbone), which runs across the front of the body, jointed at one end to the scapular spine and at the other to the top of the sternum. These attachments give the shoulder blade considerable freedom of movement. It can slide up and down the back, which is what happens when we shrug our shoulders. You can feel the movement of the collarbone that this involves, by placing your other hand over it.

The shoulder blade is also free to rotate around its attachment to the collarbone, an action that supplements the movement at the shoulder joint. Starting with your arm hanging down by your side, swing it outwards and then upwards. You will not be able to make the arm point vertically upwards unless you bend your spine to the side. If you keep your spine straight you will probably be able to swing your arm through about 150 degrees, of which about 90 degrees will be rotation of the shoulder joint proper, between the humerus and the shoulder blade. The remaining 60 degrees will be rotation of the shoulder blade.

The deltoid muscle, which makes a prominent bulge on the shoulder, runs from the collarbone and scapular spine to the humerus. Other muscles run to the humerus from both the outer and the inner face of the shoulder blade. These muscles are important not only for moving the shoulder joint but also for holding it together, preventing

*Plate 68 (opposite) An adult pelvic bone. There is little sign of its having developed from three separate bones. (Courtesy Maxilla & Mandible.)*

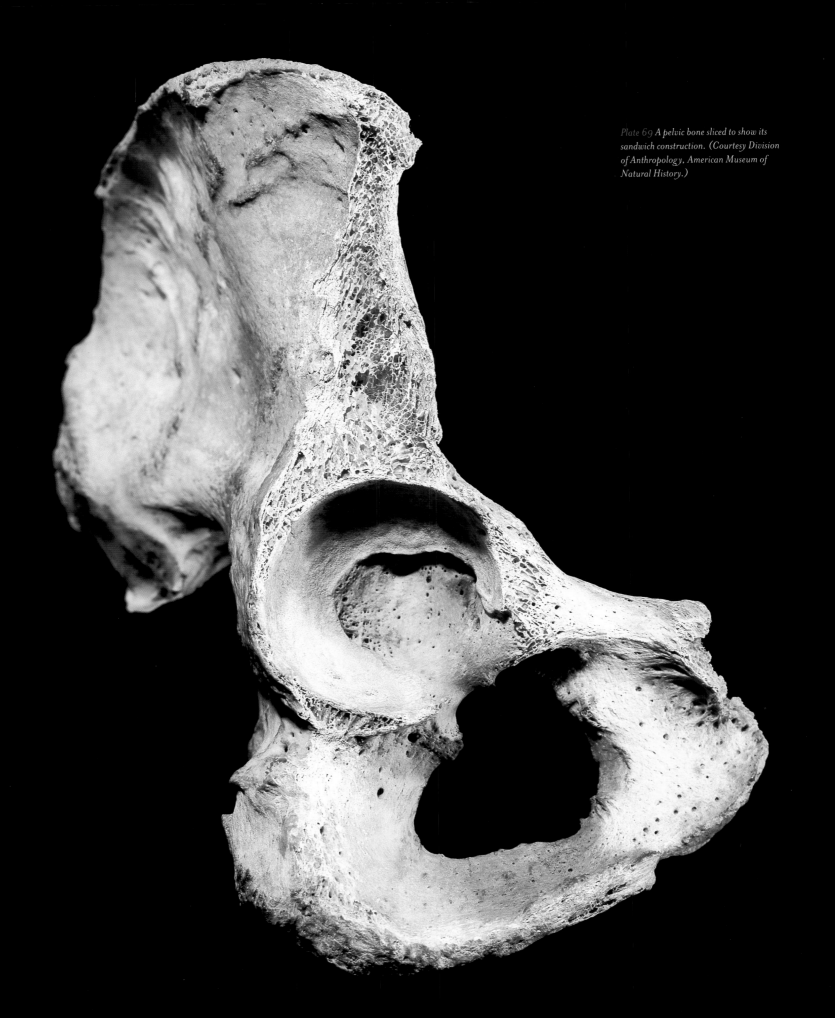

Plate 69 *A pelvic bone sliced to show its sandwich construction. (Courtesy Division of Anthropology, American Museum of Natural History.)*

dislocation. Most of the muscles that move the shoulder blade and hold it in place on the rib cage attach to the edge of the blade (the edge nearest the backbone).

## PELVIC BONES

The skeletons of human and animal hips are astonishingly different from the skeleton of the shoulders. Legs are built on the same basic plan as arms, as we saw in Chapter 3. Despite this, the pelvic bones (with which the femora articulate) are very different from the shoulder blades. Evolution has solved the problem, of attaching a limb to the torso, in two very different ways. At the shoulders, we have shoulder blades loosely attached to the ribs by muscles. At the hips we have left and right pelvic bones firmly attached to each other and to the sacrum to form a very strong, relatively rigid ring. [Plate 67] It is not quite rigid because the sacroiliac joints, between the sacrum and the pelvic bones, allow a few millimeters of sliding and one or two degrees of rotation. This slight mobility seems to have no importance in human movement. The sacroiliac joints are principally notable because they seem to be responsible for a great deal of back pain.

In babies, each side of the pelvis consists of three distinct bones connected by cartilage. It is only after puberty that these three bones fuse to such an extent that the boundaries between them disappear. [Plate 68] The uppermost of these three bones is the ilium, which forms the broad blade of bone above the hip joint. Its prominent upper edge can be felt just below the waist. The lower parts of the pelvic bone, below the hip joint, are formed by the ischium (behind) and the pubis (in front). The bony lump in the buttocks, that can make sitting on a hard chair uncomfortable, is the corner of the ischium. Like the bones of the skull roof, the pelvic bone consists of two layers of compact bone with spongy bone sandwiched between them. [Plate 69]

Like the shoulder blades, the pelvic bones have large areas for muscle attachment. The erector spinae, the big muscle that straightens the back, attaches to its upper edges (and also to the sacrum). The muscles of the wall of the abdomen, which people who value their trim figures take such trouble to exercise, also attach to the upper edges of the pelvic bone. The muscles of the buttocks attach to the sacrum and to the outer face of the ilium. Most of the muscles of the thigh attach lower down, on the outer faces of the ischium and pubis.

Like previous chapters, this one has shown that the skeleton must be understood partly as the product of design by natural selection, and partly by the accidents of ancestry. Evolution has given us ribs well designed for breathing, and a backbone with joints that give the flexibility we need. But the coccyx seems to be the useless relic of a monkey-like tail, and the sacrum has been cobbled together from pre-existing vertebrae and ribs. The totally different designs of our shoulders and hips work well for us, but have their origins in the requirements of fore and hind fins in the fishes from which we evolved.

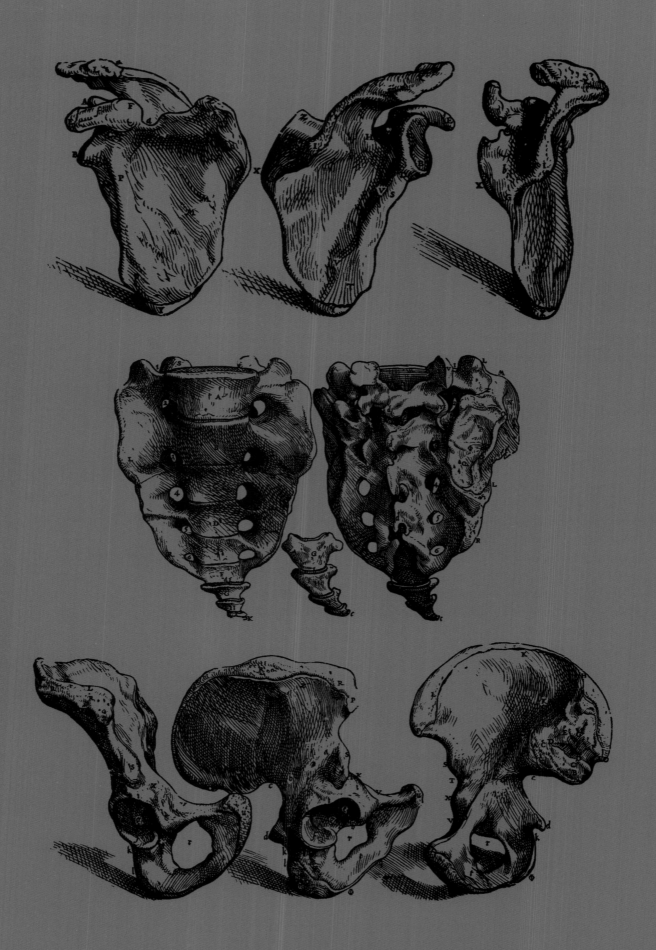

# 5

# DISEASED & DAMAGED BONES

So far in my life, I have been lucky. None of my bones have been broken. Medical records show that British people have one chance in 2500 of breaking one or other humerus in the course of a year, and one chance in 1600 of breaking one of the two femurs. For a lifetime of (say) 70 years, that implies a 1.4% chance of breaking each humerus, and a 2% chance of breaking each femur. Those percentages may seem low, but we have a lot of bones in our bodies, so the chance of breaking at least one bone in a lifetime is quite high.

Many of these bone fractures are due to special hazards of modern life, such as traffic accidents. The hazards were different in the distant past, but the probabilities of bone fractures may have been quite similar. For example, archaeologists investigating an ancient Native American site in Ohio found healed fractures in 72 of the 2383 bones that they examined; that is, in 3% of the bones. They counted only healed fractures to avoid including bones that had got broken after death. This implies that they failed to count people who died of injuries that broke their bones. The modern medical records excluded people who had died in the accidents that caused the fractures (dead people do not consult physicians), so the two sets of data are comparable.

As explained in Chapter 3, long thin structures such as bones are at risk from forces that bend them. Bones broken in bending generally have fractures that run transversely across the shaft, more or less perpendicular to the long axis of the bone. Bones also get broken by twisting, which causes spiral fractures that wind round the bone like a spiral staircase. [Figure 14]

Plate 70 *Leg skeleton of a French soldier
injured in the First World War.
(Courtesy of Henry Galiano)*

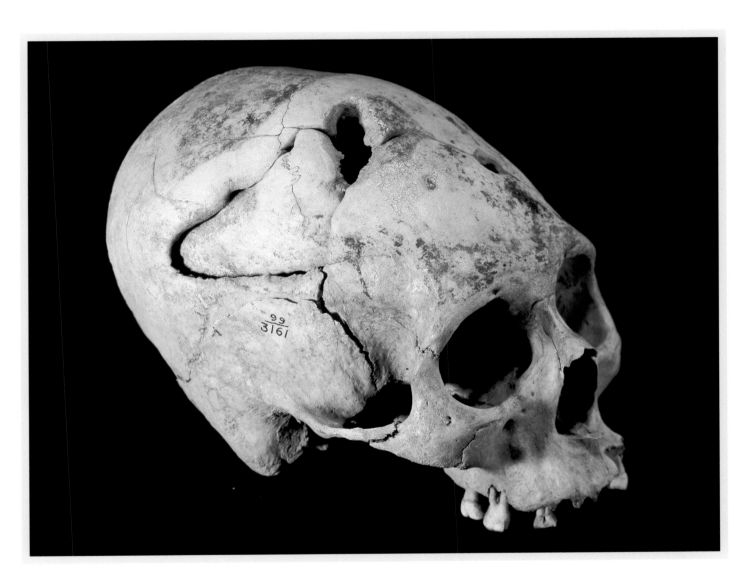

PLATE 71
A skull from Peru, with a seriously damaged skull. New
bone growth along the edges of the wound shows that this
victim survived for long enough after being injured for
repair processes to start. (Courtesy Division of
Anthropology, American Museum of Natural History.)

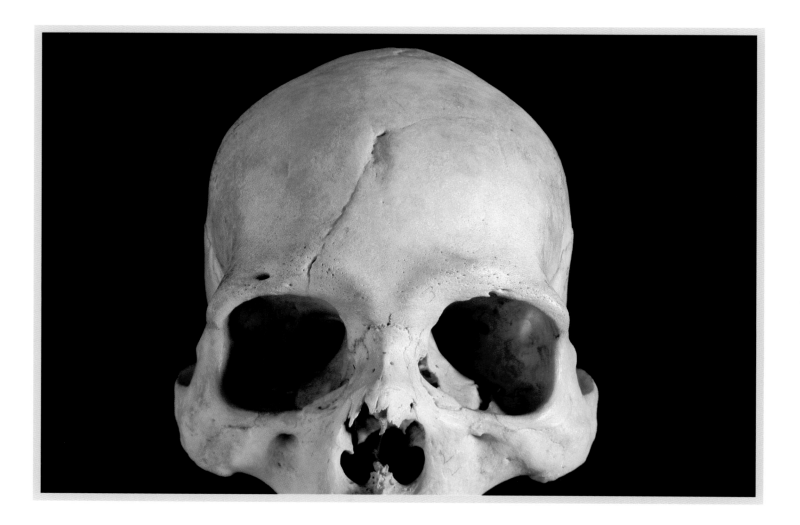

## MECHANICS OF BONE BREAKAGE

We have no precise knowledge of the mechanics of bone-breaking accidents. We can set up experiments to make accurate measurements of the forces involved in activities such as running and jumping, but few people would volunteer to be subjects in experiments on bone fracture. Though we cannot measure the forces, it is often fairly clear what has happened in accidents captured on video or film.

In many situations, the forces that can act on bones are limited by the strength of muscles. A large force on my shin will generally stretch my knee muscles and bend my knee, moving my tibia out of the way before the force has grown large enough to break it. If, however, an external object prevents my knee from bending, larger forces can act. Figure 15 shows an accident (or possibly a deliberate foul) in soccer. The victim's foot is on the ground when his opponent's foot hits his shin. The lower end of the victim's tibia is held in place by the foot on the ground, and the upper end by the inertia of the

*Plate 72 A skull with a healed fracture running diagonally across the forehead. (Courtesy Division of Anthropology, American Museum of Natural History.)*

victim's body. In this situation, very large bending forces may act, causing transverse fracture of the tibia and fibula. Spiral fractures may occur in skiing, when one ski is suddenly deflected by an obstacle, twisting the leg. [Figure 16]

We want our bones to be strong enough not to break, but we do not want them too heavy. If our leg bones were twice as thick as they actually are, we would be less likely to break them, but we could not run so fast. The ideal must be a compromise between strength and lightness, but what is the best compromise? Plainly, the bones must be at least as strong as is needed for everyday activities.

The stresses that act in bones in life can be investigated in various ways. The least indirect method uses strain gauges. These are small pieces of metal foil or semiconductor material, bonded onto paper. Their electrical resistance changes when they are stretched or compressed. For one set of experiments, the surface of a brave volunteer's tibia was exposed surgically, and strain gauges were glued to it. They were connected to recording instruments by wires brought out through his skin. Stresses in the tibia stretched or compressed the gauges. As the elastic properties of bone are known, the stresses could be calculated from the output of the gauges. Records made while he ran slowly showed peak stresses that were only about 7% of the stresses that would be needed to break the bone. It is not clear, however, whether the gauges were placed on the most highly stressed part of the bone, and in any case larger stresses would have acted in faster running. In a similar experiment on horses jumping over a bar, stresses up to one third of the breaking stress were recorded. Thus the safety factor for these (quite small) jumps was about three. Experiments on other animal species have shown that many bones have factors of safety between two and five for activities such as running and jumping.

Those factors of safety allow some latitude for things to go wrong. Larger stresses may act in a fall or other accident, than in normal running or jumping, but bones are unlikely to be broken unless the accident is a severe one. Also, the factors of safety help to ensure that bone fatigue does not build up too fast. We saw in Chapter 1 that quite small stresses may eventually lead to fracture if they are repeated too frequently. Bone cells repair fatigue damage, but can do so only at a limited rate.

## REPAIR OF BROKEN BONES

Broken bones heal. The broken ends bleed, and a blood clot forms around them. New blood capillaries develop and grow into the clot, bringing white blood cells that clear away the debris of damaged tissue. Within a few days, a cuff of collagen fibers forms around the break, binding the parts of the broken bone together. After a few weeks, hydroxyapatite is deposited in this cuff, converting it to a bony callus. Eventually,

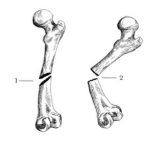

*Figure 14 Bone breakage.*
*(1) Spiral fracture, (2) Transverse fracture*

*Figure 15 A soccer accident that may cause a transverse fracture.*

*Figure 16 A skiing accident that many cause a spiral fracture.*

typically about three months after the accident, the two ends become firmly united. Children's bones heal faster.

The callus that effects the repair is a substantial lump of bone in which the collagen fibers run irregularly in all directions. A remodeling process that continues for several years reduces the size of this lump and replaces the irregular new bone in the break with neatly arranged lamellae and osteons. [see Figure 2] The repair remains detectable, however, in bones examined after death.

Healed fractures in wild animals show that healing is possible without medical attention. A bone cannot heal properly, however, if the broken edges are left out of alignment. Plate 70 shows, among other injuries, an imperfectly healed fracture of the tibia. Plate 71 shows a fractured skull with the broken edges too far apart to heal.

Many fractures of arm and leg bones can be kept in alignment during the healing process by a plaster cast, which also protects the healing bone from injury. In other cases, it may be necessary to fix the parts of the broken bone together with screws, or by means of a metal plate attached by screws.

## AXES, SWORDS, AND BULLETS

Different weapons damage bones in different ways. The effects depend on the velocity of impact and the sharpness of the weapon or missile. Blunt weapons such as clubs, hitting a head at relatively low velocity, may produce a few long cracks running across the skull. Alternatively, they may cause spider's web fractures, with some cracks radiating out from the point of impact and others forming rings round it. Sharp axes and swords cut into the skull. [Plate 72-73] A bullet, traveling at much higher velocity, punches a clean hole. [Plate 74]

Any cracks around the hole are generally short, presumably because the force of the impact was too brief for cracks to spread far. Bullets travel at supersonic speeds, faster than cracks can propagate through bone, so we cannot expect more than a few millimeters of crack to form while a bullet is passing through. Viewed from the outside, the hole where a bullet entered the skull has a diameter little larger than the bullet. On the inner side of the skull, a crater of considerably larger diameter surrounds the hole. Similarly, windows struck from outside by air gun pellets may show a crater on the inner face of the glass. The hole where a bullet left the skull after going right through the head has its crater on the outside. The craters, with walls sloping at about 45 degrees to the bullet's path, are presumably due to compressive stresses in the shock wave generated by the impact. When solids are broken by compression, they commonly fail by shearing along planes at 45 degrees to the direction of the applied force.

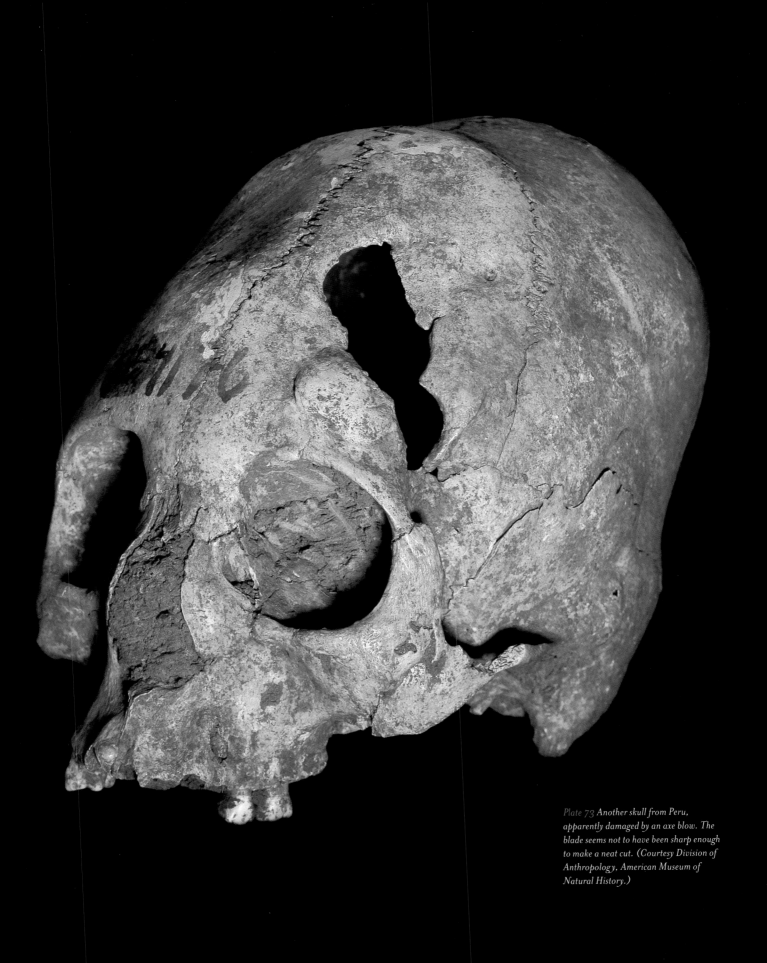

Plate 73 *Another skull from Peru, apparently damaged by an axe blow. The blade seems not to have been sharp enough to make a neat cut. (Courtesy Division of Anthropology, American Museum of Natural History.)*

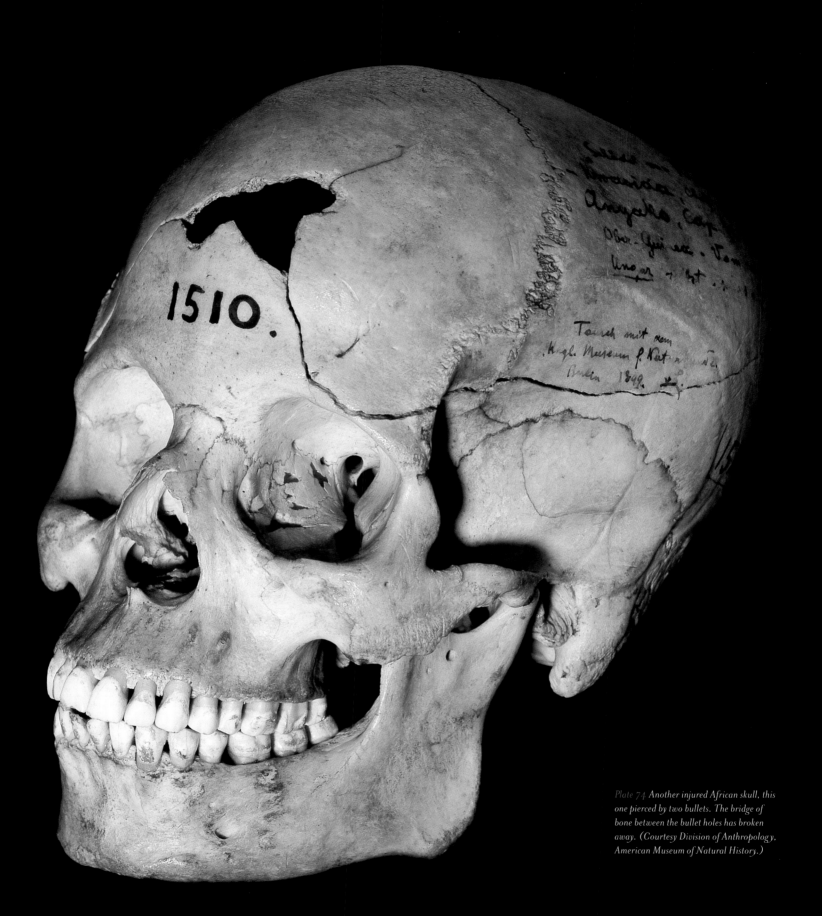

*Plate 74 Another injured African skull, this one pierced by two bullets. The bridge of bone between the bullet holes has broken away. (Courtesy Division of Anthropology, American Museum of Natural History.)*

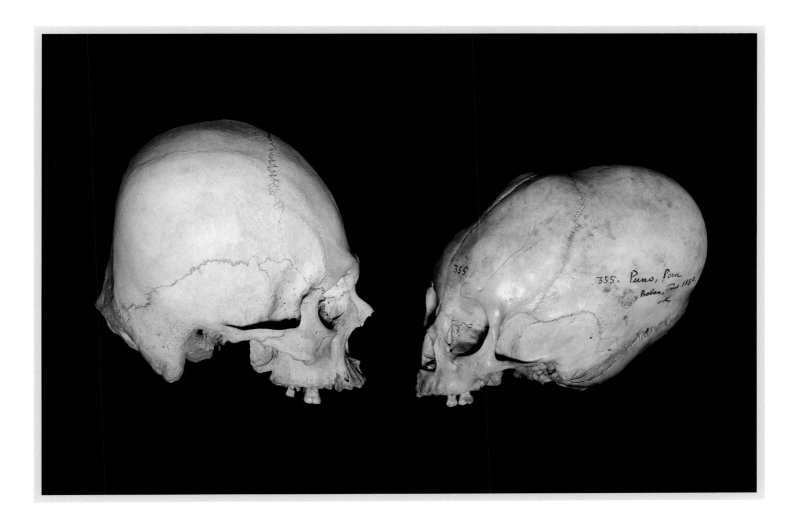

## BONE DEFORMATION AND DISEASE

In some cultures it has been customary to deform children's bones deliberately. Some American tribes (in both North and South America) used to bind babies' heads to cradleboards, flattening the back of the skull. [Plate 75] Alternatively, in some other American cultures and in Melanesia, babies' heads were bound tightly with bandages, forcing the skull into an elongated shape that was presumably considered attractive. These treatments would not have much effect on an adult skull, but until the bones become rigid and form firm sutures a baby's skull is more easily molded. Chinese girls' feet used to be bound, stunting their growth and exaggerating the arch of the foot. [Plate 76]

Bones also get damaged and deformed by disease. Wear and tear on joints causes osteoarthritis. Most sufferers are middle-aged or elderly people, but intensive sport can bring the condition on early. Tennis and other racquet games may result in osteoarthritis of the shoulder and elbow. Footballers get osteoarthritis of the knee.

*Plate 75 Skulls deformed in infancy, by binding to cradleboards. (Courtesy Division of Anthropology, American Museum of Natural History.)*

There are also some jobs that make workers particularly liable to osteoarthritis. In a study published in 1978, clinicians examined the hands of 64 women who had been working for 20 years or more in a textile mill in Virginia. They were still at work, but many of them showed symptoms of osteoarthritis. Some of them were winders, doing a job that involved a lot of wrist movement but little movement of the fingers. More of them than of the other workers had osteoarthritis in their wrists. Others were burlers and spinners, whose jobs depended more on finger movements and less on the wrist. More of them had osteoarthritis in their fingers. Though many cases of osteoarthritis can be explained as the results of sports or occupations others, especially in women over the age of 50, have no apparent cause.

In joints damaged by osteoarthritis, the cartilage that is so important for lubrication gets thinned or even worn away. The joint becomes stiff and painful. If the cartilage is lost, bones rubbing directly on each other may wear each other down. Also, outgrowths may develop from the bones around the joint. [Plate 77] Rheumatoid arthritis produces symptoms that are in some respects like osteoarthritis. Rather than outgrowths forming on the bones, however, bone near the joints gets eaten away. More fundamentally, the cause of this disease is entirely different from that of osteoarthritis. The cartilage is not worn away, but destroyed by malfunction of the body's immune system. The disease usually appears first in the hands, and spreads to other joints.

Joints that are severely damaged either by osteoarthritis or by rheumatoid arthritis are often replaced with prostheses. One of the most successful of these has been the Charnley hip prosthesis, introduced in the 1960s. This is a ball and socket joint: a stainless steel ball that fits into a plastic (high density polyethylene) socket. The ball is mounted on a curved steel shaft. The damaged head of the femur is cut off, and the shaft of the prosthesis is fitted into the marrow cavity of the bone. The plastic socket is fixed into a hole cut in the pelvis. One of the first recipients of a Charnley prosthesis was the Marquess of Exeter, who had won the 100 meter hurdles in the 1928 Olympic Games. (At the time of the games he was known as Lord Burghley, as the previous Marquess was still alive.) His hurdling had presumably caused the damage that made the joint replacement necessary. The prosthesis improved his quality of life so much that when the first one wore out and had to be replaced he mounted it instead of the usual emblem on his Rolls Royce.

Osteoporosis is a common and troublesome bone disease. It affects mainly elderly people, women much more than men. Bone is withdrawn from the skeleton, leaving bones thinner-walled or porous, and correspondingly weak. The problem is made worse by changes that occur in the bone even of healthy people, as they age. As we get older, our bone becomes more brittle. Thus elderly sufferers from osteoporosis not only have less bone than healthy young people; the bone that remains is also more

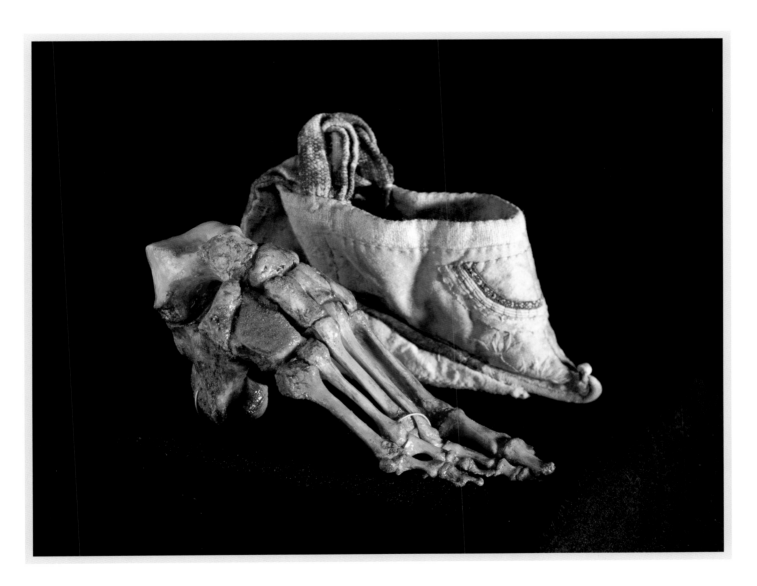

PLATE 76
*The skeleton of a Chinese woman's foot, kept unnaturally small by*
*binding during childhood, with one of her shoes.*
*(Courtesy of Deborah Galiano.)*

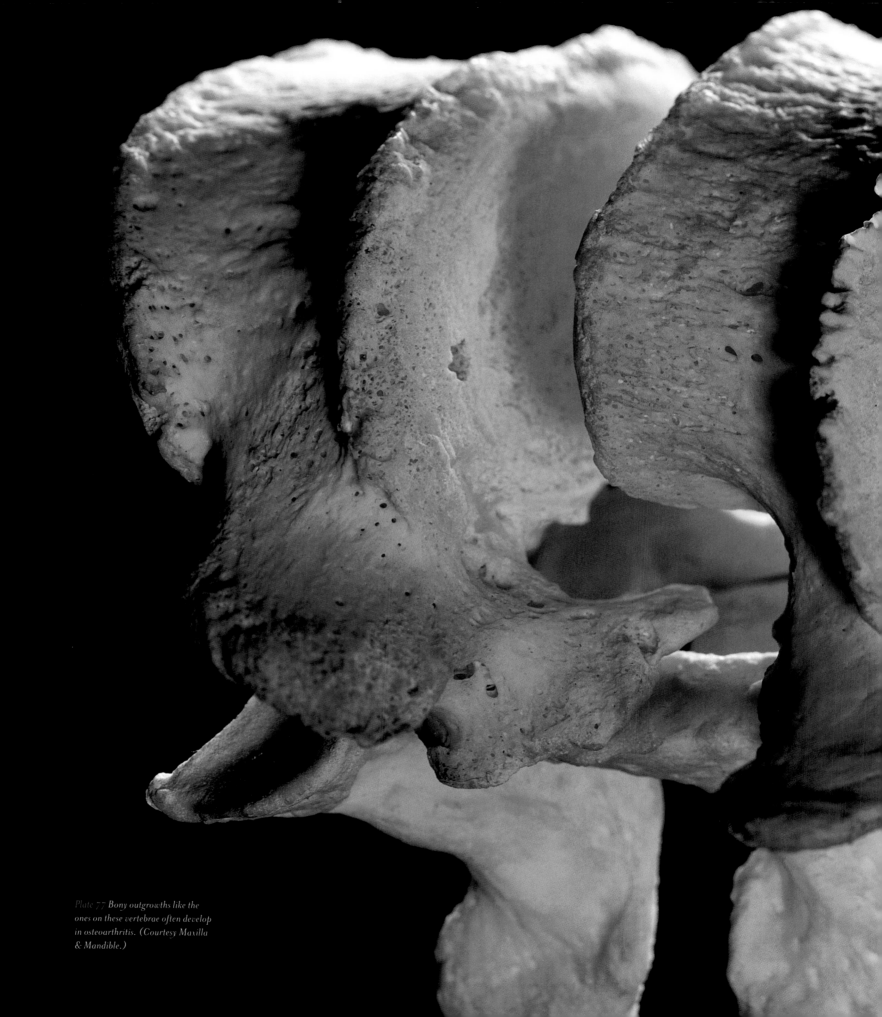

Plate 77 **Bony outgrowths like the ones on these vertebrae often develop in osteoarthritis.** (Courtesy Maxilla & Mandible.)

brittle. Falls that would do no harm to young people often break elderly people's bones. The neck of the femur is often broken if the patient falls sideways onto a hip, and the lower end of the radius is likely to be broken if she tries to break her fall with an outstretched hand.

The scope for research on this topic is strictly limited. It is not acceptable to push elderly people over to see how they fall and whether their bones break. Experiments have had to be done on students, with gymnasium mattresses to cushion their falls. A team of scientists in Boston made their subjects trip by suddenly tightening a cable attached to an ankle, as they walked through the laboratory. They simulated slipping by means of a section of floor mounted on low friction rollers, which slid forward when stepped on. Stepping unexpectedly off a kerb was simulated by means of another section of floor, which sank suddenly when stepped on. Finally, the subjects were asked to simulate fainting by allowing themselves to go suddenly limp, when a whistle blew. Not surprisingly, tripping and stepping down from a kerb resulted in forward falls, with little danger of landing on a hip. Slips and faints also resulted in forward falls when the subjects walked fast, but when they walked slowly they were more likely to fall backwards or to the side, landing on buttocks or hip. Slips and faints while walking slowly are the falls that seem most likely to result in fractures of the femur. One of the suggestions that emerged from the research was that slow moving elderly people should be encouraged to exercise, to build up the strength that would enable them to walk faster.

## BRITTLENESS

At this stage, we need to be clear what brittleness is. Roughly speaking, a brittle material is one that is easily broken by an impact. Glass and china are obvious examples. The effect of an impact depends on the energy involved. When a stone is thrown at a window, the energy available to break the glass is the kinetic energy (energy of movement) of the stone: half its mass multiplied by the square of its velocity. A heavy stone, or one thrown fast, is more likely to break the window than a small stone thrown at low velocity.

An impact deforms the colliding objects. For example, the first effect of the impact of a falling cup on the floor is that the cup gets bent slightly out of shape. As it bends, the force on the cup increases, just as the force in a rubber band increases as you stretch the rubber. The stresses in the cup increase, and if the stress anywhere comes to exceed the strength of the china, the cup will break. The energy needed to break the cup depends both on the breaking force and on the amount of deformation needed to raise the force to that level. Ideally, if the cup obeyed Hooke's Law of Elasticity (which it is unlikely to do exactly) the energy would be half the force multiplied by the deformation.

*Plate 78 (opposite) A skull affected by acromegaly. (Courtesy G.J. Sawyer.)*

That means that if you compare pieces of the same size of two materials of the same strength (that require the same stress to break them) the more flexible material will need more energy to break it, and so be less brittle. John Currey of the University of York, England, obtained the femurs of patients of various ages, who had died after short illnesses that were thought not to have affected their bones. He cut strips of bone from them, and measured the energy needed to break the strips in an impact-testing machine. He found that only one third as much energy was needed to break elderly bone, as to break samples of the same size from the bones of children. In other words, the elderly bone was three times as brittle. Currey also measured the calcium content of the bones. He found that it increased with age, and that the bones with higher calcium content were more brittle. The increase in brittleness seemed to be due to the older bones containing a bigger proportion of calcium salts (hydroxyapatite) and correspondingly less protein, and so being less flexible.

## DISEASED SKULLS AND TEETH

Plates 78, 79 and 80 show examples of diseased skulls. In Plates 78 and 79 the problem is acromegaly, a condition due to a tumor in the pituitary gland making the gland secrete too much growth hormone. Notice the thickening of the bones of the skull roof. Other symptoms of acromegaly include enlargement of the hands and feet. Plate 80 shows growths on a skull due to a genetic defect. Joseph Merrick (1862-90), who was exhibited in freak shows as "The Elephant Man", suffered from a similar condition. Plate 81 shows a bad case of bones, which should have remained separate, having sutured together. The lower jaw is immovably fixed to the skull, so food must have been given by tube through the gap that has been made in the front teeth. In addition, the cervical vertebrae have fused together, making the neck rigid.

Dental caries (tooth decay) is a disease that gives many of us a lot of trouble, and dentists a lot of employment. [Plate 82] The trouble starts in bacterial dental plaque, which is a soft, sticky, transparent film that develops on our teeth. It is a mixture of saliva, bacteria and food debris. When we eat, more food debris gets caught in the plaque. The bacteria break down the sugars in this food, producing acids. These dissolve the hydroxyapatite in enamel and dentine, and eat into the teeth. The plaque remains acid enough to cause damage for around 20 minutes, before the acid is neutralized by the saliva.

The danger of caries depends less on how much sugar we eat, than on how often we eat it. There is an opportunity for acid to attack the teeth after every sugary meal or snack. The effects of sugary foods were clearly shown by an experiment in the 1940s, in which some children were given toffees to eat between meals, and others were not. The children who got toffees may have thought themselves lucky, but they suffered more tooth decay.

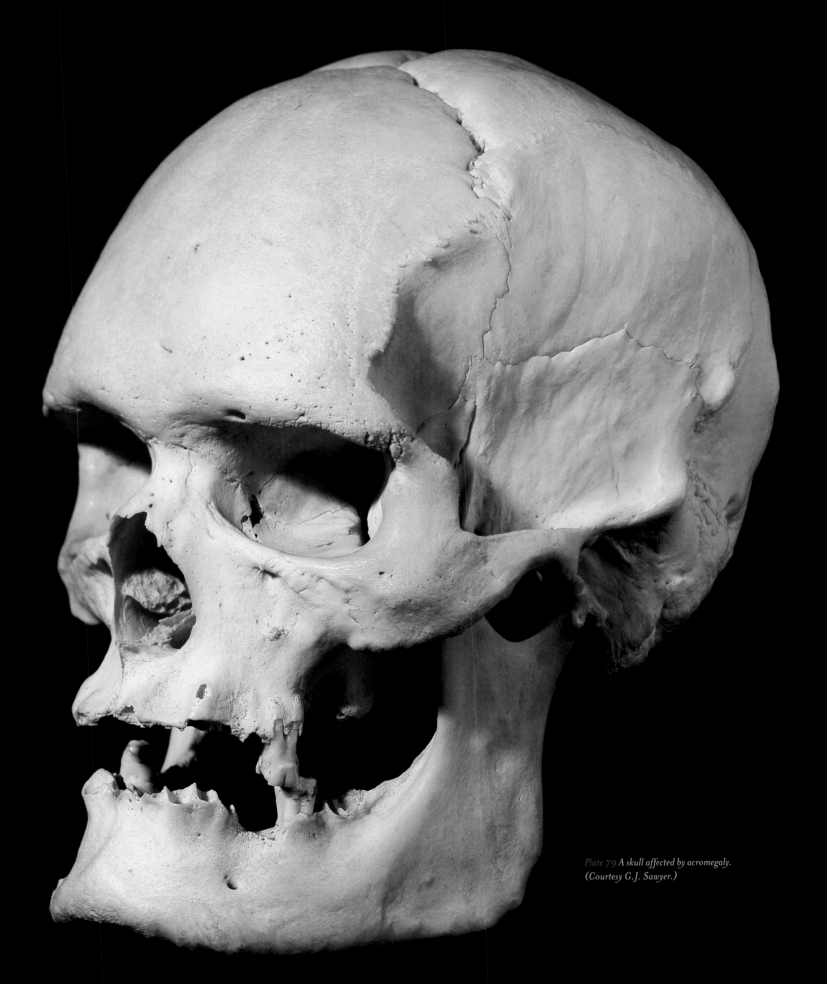

*Plate 79 A skull affected by acromegaly.*
*(Courtesy G.J. Sawyer.)*

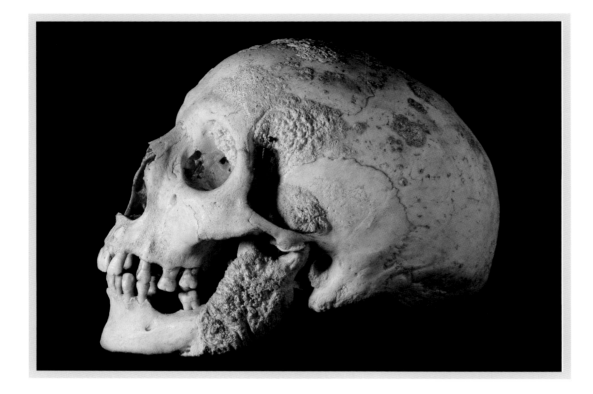

Plate 80 *A skull affected by a genetic disorder, similar to that of the Elephant Man. (Courtesy Henry Galiano.)*

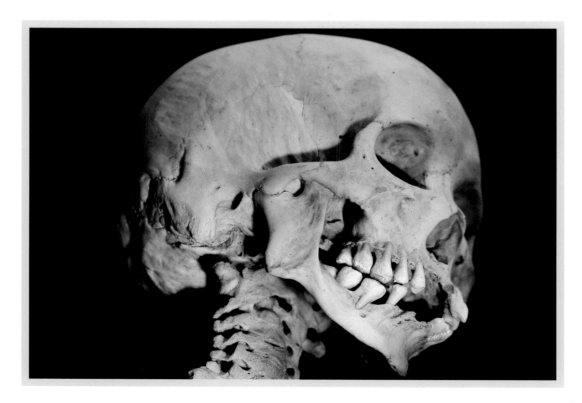

Plate 81 *The lower jaw of this skull has become rigidly sutured to the skull, and the cervical vertebrae are fused together. (Courtesy Henry Galiano.)*

Food fragments are most likely to accumulate in the gaps between one tooth and the next, and in the fissures between the cusps of teeth. These places are especially liable to decay. Brushing the teeth after meals can remove a lot of the food fragments and plaque, but generally fails to clean every crevice. Fluoride in the drinking water or in toothpaste considerably reduces the incidence of caries.

Caries is especially prevalent in children. Tooth loss in older people is more often due to periodontal disease. This also is caused by bacteria in plaque, but in this case the foods in the plaque do not have to contain sugars. Toxins produced by the bacteria cause inflammation of the gums. In bad cases, some of the bone of the tooth socket may be destroyed, loosening the tooth. Thorough and regular cleaning of the teeth will generally prevent the disease.

There was remarkably little caries, by modern standards, in the teeth of the ancient Romans who were buried at Herculaneum by the eruption of Vesuvius in 79 AD. Mediaeval European teeth also generally show relatively little of the disease. Several excavations of mediaeval cemeteries have found that only 5-11% of the teeth were affected. Much more caries was found in skeletons recovered from the wreck of the Mary Rose, a warship which sank in 1545, and two investigations of eighteenth century burials found caries in about 18% of the teeth. Sugar did not become widely available until sugar cane plantations were established in the West Indies in the mid seventeenth century, but caries seems to have increased before that happened. The explanation may be that flour was being ground more finely. Bread made with coarse mediaeval flour was less likely to stick to the teeth than bread made with the finer, later flour, and abrasive particles in the coarse flour would have helped to scour plaque off the teeth.

Decayed teeth may benefit from repair, and people who lose teeth may want replacements either for use or for cosmetic reasons. False teeth have been made at least since Roman times. A skull of about 100 AD, found in France, has an iron premolar pushed into the socket of the lost tooth. For many centuries after that, restorative dentistry was only for the privileged. The 968 skeletons in the crypt of a church at Spitalfields in London had to be moved from the crypt, which was needed for other purposes. Many of them had plates attached to their coffins, giving details of the occupants. They had died between 1729 and 1852. Only 12 of them had had teeth repaired or replaced. Among them, Deborah Peck who died in 1739 had a piece of ivory, carved to represent two missing incisors, tied with silk to the adjacent teeth on either side. She was a wealthy widow. Eliza Favenc (died 1809), wife of the Consul for the Canary Islands, had eight fillings made by stuffing crumpled gold foil into cavities. William Laschollas, a stationer who shot himself in 1852, had a gold upper denture with four porcelain teeth.

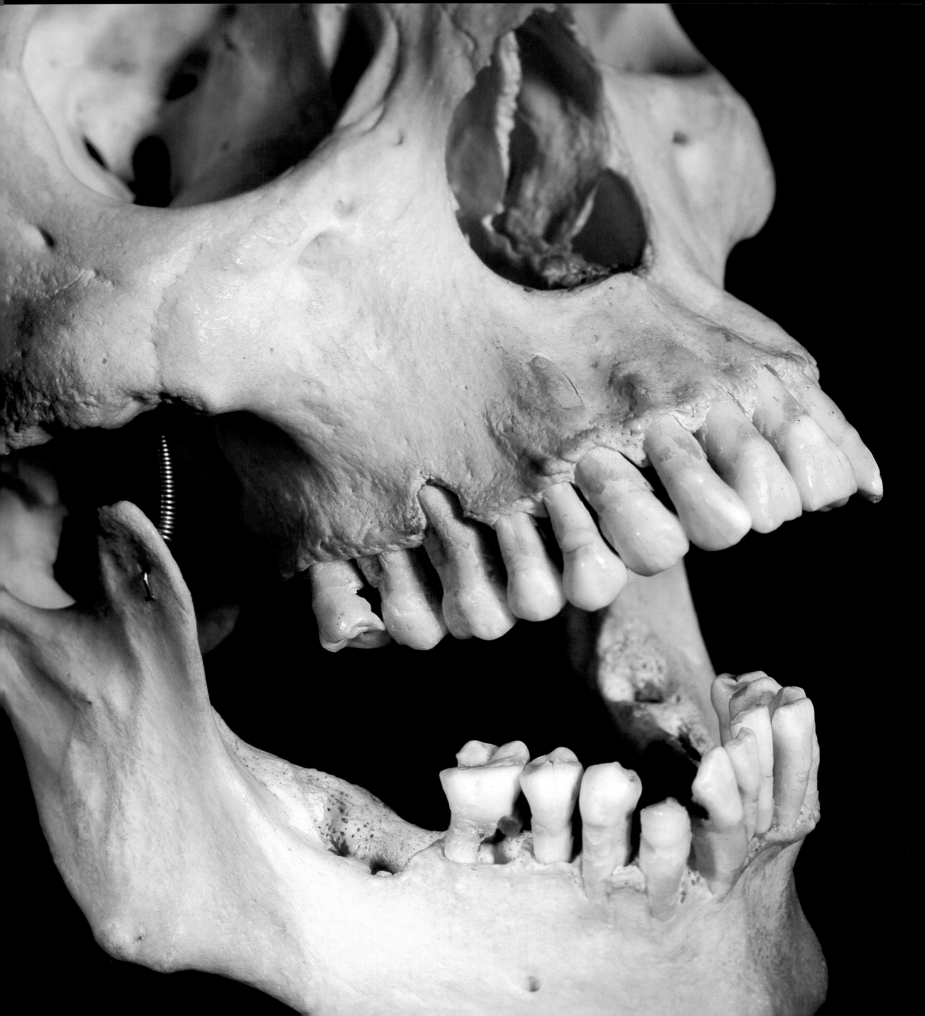

Amalgam has long been the commonest filling material. An alloy of silver, copper and tin, in powdered form, is mixed with mercury. The resulting paste is forced into the tooth cavity and shaped as required. The mercury gradually dissolves the metal powder, producing a solid amalgam, which becomes durable enough for chewing within a few hours. Amalgam fillings wear very well, but would be conspicuous if they were used at the front of the mouth. Composite fillings are an alternative that can be made to match the color of the teeth. They consist of a plastic resin mixed with a ceramic powder (for strength and color) and a catalyst. In the presence of the catalyst, the resin polymerises and solidifies. In older systems, the catalyst was mixed in immediately before use, as in glues that require mixing material from two tubes. More recently, light-activated catalysts have come into use. The catalyst can be mixed in by the manufacturer, and polymerization does not occur until the dentist shines a bright light on the completed filling. Neither amalgam nor composite fillings adhere to the teeth. The cavity has to be undercut, so that the deeper parts are wider than the entrance and the filling cannot fall out. Crowns, however, have to be fixed with adhesive cement. They are made of porcelain, with or without a gold backing.

In this chapter we have seen some of the ways in which bones and teeth get damaged by violence and disease. In the next we will be concerned with natural differences between different people's bones.

*Plate 82 (opposite) Teeth have decayed and been lost from this neglected mouth. (Courtesy Maxilla & Mandible.)*

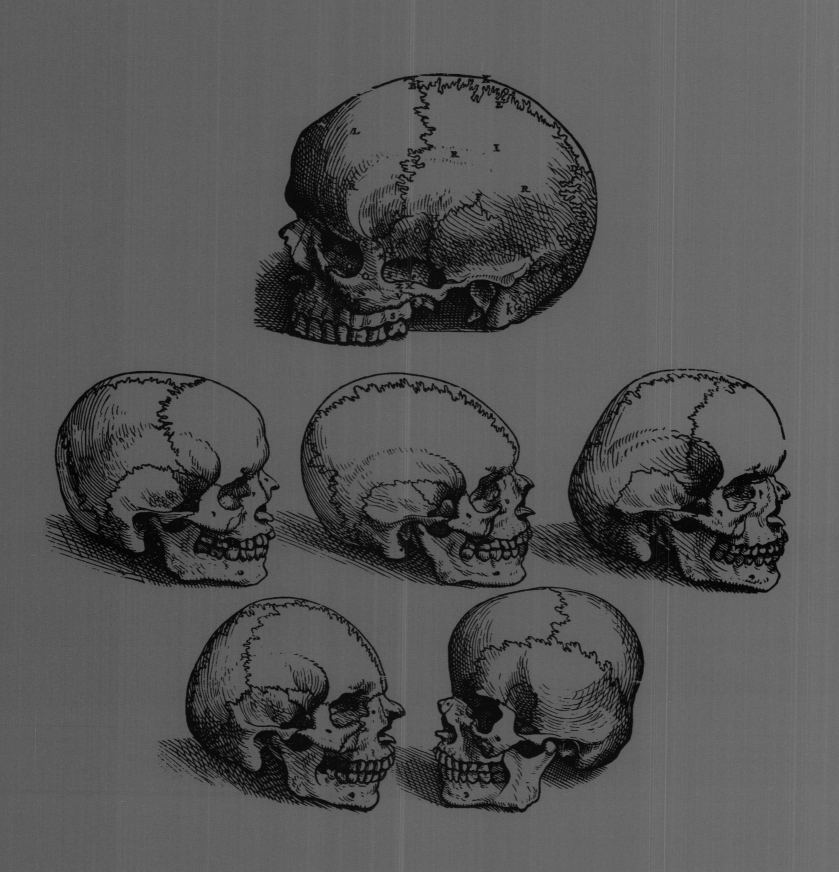

# 6

# DIFFERENT PEOPLE'S BONES

Everybody's skull is different. If you look carefully at the three adult male European skulls in Plate 83, you will find many small differences between them. Similarly in Plate 84 the skulls are different shapes, and the sutures between the bones show different patterns of wiggles. In the skull on the left, there is a clear suture down the middle of the forehead, but in the one on the right (as in the great majority of adults) the bones of the two sides of the forehead are fused. If we were all reduced to skeletons and walked around like so many ghosts, we could probably recognize each other's skulls as easily as we now recognize each other's faces. [Plate 85] It would take us some time to become as familiar with the details of skulls as we are with faces, and to learn about the kinds of variation to look out for, but I am confident that we would soon learn to distinguish our friends.

Differences between individuals can also be found in other bones. For example, the femur on the right in Plate 86 has a stouter, straighter shaft than the others. The forearm bones on the left, in Plate 87, are shorter than the ones on the right and a little more robust.

## FORENSICS

Forensic scientists study the differences between people's skeletons. If a skeleton is found, the police want to know who it is. Clothing or jewelry found with the skeleton may identify it. A pattern of dental fillings, crowns and missing teeth can often be matched with dental records. Healed fractures in bones can sometimes be matched with hospital records, which may include X-radiographs. Alternatively, it may be possible to extract samples of DNA from the skeleton and compare them with the DNA of a missing person, or to compare them with samples from the missing person's family.

At any one time a great many people are missing. Many of them may not have been reported as missing to the police. For this reason, trying to identify a skeleton by means of dental records or DNA may be like looking for a needle in the proverbial haystack. It is very helpful to be able to reduce the range of possibilities. If a forensic scientist can tell the police, for example, that a skeleton is that of a tall black woman aged about 45, they can limit their investigation to people of that description.

## MALE VS. FEMALE

There are fairly consistent differences between male and female skulls, but they are undeveloped in children and modified in old people. They work best between the ages of 20 and 55. Male skulls are generally larger than female ones, but there is considerable overlap in size. More reliable indicators are the eyebrow ridges over the eye sockets, which are more marked in men; and the larger, squarer lower jaws of men, with more prominent chins. [Plate 88] The mastoid process, which projects from the skull close behind the ear, is smaller in women. These are not clear-cut differences, but matters of degree, so assignment of sex depends critically on the experience and judgment of the pathologist, who has to consider many features. Attempts have been made to make it more objective by using a complicated formula based on precise measurements of skull dimensions. This demands much more work than simply looking at a skull and forming an opinion, and seems to be no more reliable. In tests on skulls of known sex the formula gave the correct answer in only 85% of cases. The differences between the skulls of the two sexes are less marked for some races than for others, making determination of sex more difficult. It is particularly difficult for people whose ancestors came from the Indian subcontinent.

The pelvic girdle seems to be a more reliable indicator of sex than the skull. With it, experienced pathologists can identify sex correctly in 95% of adult cases. Wider hips and other small differences combine to give the female pelvis a relatively larger birth passage. [Plate 89] Women who have had children may have scars on the pelvic bones where the attached tissues have been torn by the trauma of childbirth. The lower edge of the pelvis (immediately above the vulva or penis) has a broader notch in women than in men. On either side of that, the apertures in the lower parts of the pelvic bones are more triangular in women and more oval in men.

Other bones are less helpful than the skull and pelvis, for distinguishing the sexes. Men generally have longer, more robust humeri and femurs than women, with larger heads. [Plate 90] A very long femur is almost certainly male. There is a difference

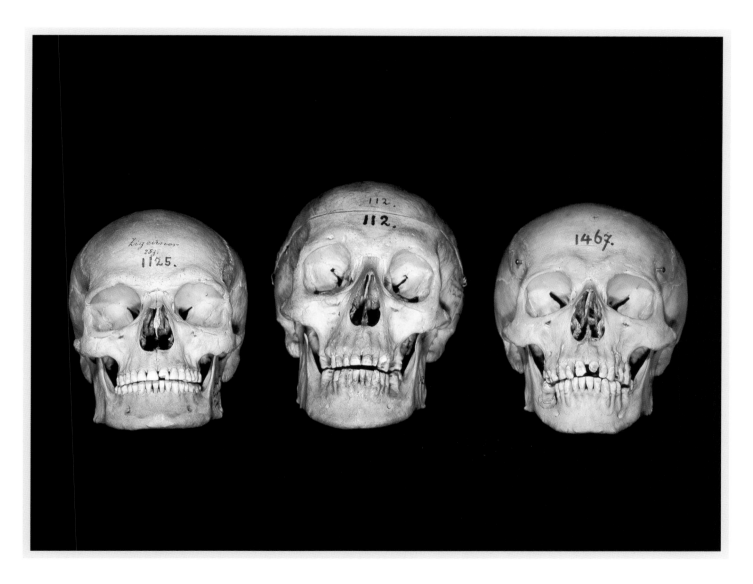

PLATE 83
*No two skulls are identical. These are all adult European men.*
*(Courtesy Division of Anthropology, American Museum*
*of Natural History.)*

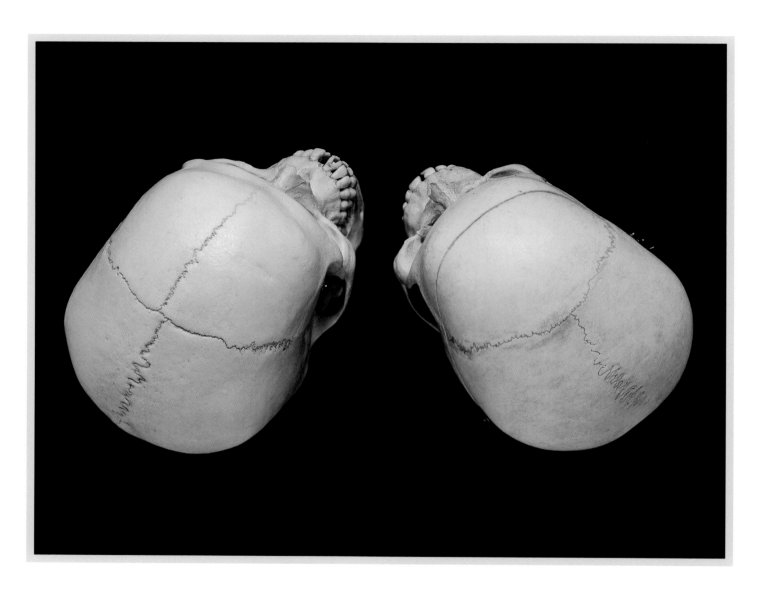

PLATE 84

*The skull on the left retains a suture that has fused in the
one on the right. (Courtesy Division of Anthropology,
American Museum of Natural History.)*

in the shape of the lower end of the femur that can be explained by the wider hips of women. We stand with our knees closer together than our hips. Consequently the femur slopes in towards the mid line, whereas the tibia is vertical. Because women's hips are wider, their femurs slope in at larger angles and meet the tibia at a larger angle, so the surfaces that articulate with the tibia have to be set at a different angle. The difference is small, only about four degrees, but it can be measured. The other limb bones, like the femur, are generally larger in men than in women, but are less useful for distinguishing the sexes. The sternum tends to be shorter in women than in men.

## STATURE

Stature is relatively easy to estimate if a full skeleton is available, by arranging the bones in their correct relative positions and measuring the assembly. Care must be taken to leave appropriate gaps between bones for the articular cartilage and intervertebral discs, which are present in the living body but not in dry skeletons. Allowance must also be made for the thickness of the scalp and of the fatty pads that cushion our heels. Even when the skeleton has been assembled and measured carefully, the estimated height may be wrong by as much as 5 cm.

Stature can also be estimated from individual bones. The femur is the best, with the tibia the second choice. Many formulae have been worked out for calculating stature from measurements of different bones and combinations of bones. Different formulae are needed for the two sexes and for different races, if reasonably accurate results are to be obtained. One of the best sets of formulae is based on measurements of casualties from World War II and the Korean War, and applies only to young adult males. Even with a wide choice of formulae available to suit different kinds of people, individual differences in body proportions make errors inevitable. For example, I am tall but my back is not unusually long. I seem to owe my height principally to very long femurs. As a result, I am excruciatingly uncomfortable in closely spaced theatre seats, but do not obscure the view of the person sitting behind me too badly.

## AGE

It is easier to estimate the ages of young people than of old ones. In infants, age can be estimated by checking which epiphyses have formed. For example, the bony epiphysis on the lower end of the femur is generally present at birth, but the head of the femur is still cartilaginous at birth and does not become bone until the third month. As well as the head, which articulates with the pelvic bone, the upper end of the femur has another epiphysis where some of the hip muscles attach. That epiphysis does not ossify until the child is two years old in white girls, or three in boys. The carpal bones of the wrist ossify in turn at predictable ages over the first three years of life.

Epiphyses become useful again for estimating age, in older children and young adults. In this age range, the cartilaginous epiphysial plates become bone, so that epiphyses and shaft become fused as a single continuous piece of bone. [see Plate 8] Conveniently for forensic pathologists, different epiphyses fuse at different ages. For example, the epiphysis at the lower end of the humerus generally fuses between the ages of 13 and 16 years, the one at the upper end of the same bone between 16 and 23, and the epiphysis on the clavicle where it meets the sternum (generally the last epiphysis to fuse) between 23 and 28.

After the fusion of the last epiphysis, indicators of age are harder to find. One of the best is the texture of the pelvic bones, near the suture where they meet. As age advances, the bone there loses a pattern of ridges and grooves, becoming smoother and grainy. The method is subjective, but it is claimed that experienced pathologists can judge the ages of pelvises within a range of about six years. Another method that can be used even on fragments of bone depends on making thin sections of the bone and examining it microscopically. Young people's limb bones consist partly of layers of bone arranged concentrically around blood vessels to form osteons [see Figure 2], and partly of layers parallel to the bone surface. As people get older, the proportion of osteons increases.

Teeth as well as bones are useful indicators of age. Severe tooth wear and tooth loss suggest old age, but depend also on diet and dental care. [Plate 91] The ages of children can be estimated by observing which teeth have erupted, but there is a long interval when nothing changes between eruption of the second adult molars at an age of about twelve and eruption of the third ones at about eighteen. The roots of teeth continue to grow after the tooth has erupted, and their initially pointed tips become rounded. These changes are most easily seen in X-radiographs.

Sections made by grinding through the teeth of infants reveal a line that appears at the time of birth, and fainter striations that are formed at intervals of about a week while the tooth is being formed. Age can be estimated by counting the striations that have been formed after the birth line. In adults, there are changes in the dentine that can be seen in sections. More translucent dentine appears at the tip of the tooth and progresses towards the root, at a rate that is predictable enough to give estimates of age within about six years.

Changes also occur in amino acids, the units from which proteins are built. Amino acids can exist in two forms that are mirror images of one another. The left-handed forms are used for synthesizing proteins, but as teeth age some of the left-handed amino acids change to the right-handed form. Age can be estimated from the ratio of the two forms, provided allowance is made for the process continuing after death.

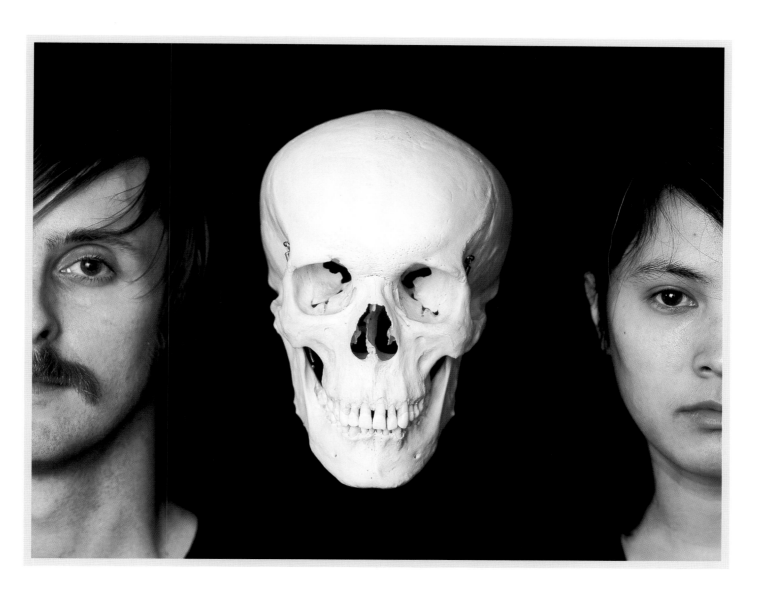

PLATE 85
*Skulls, like faces, reflect individual differences.*
*(Courtesy Maxilla & Mandible.)*

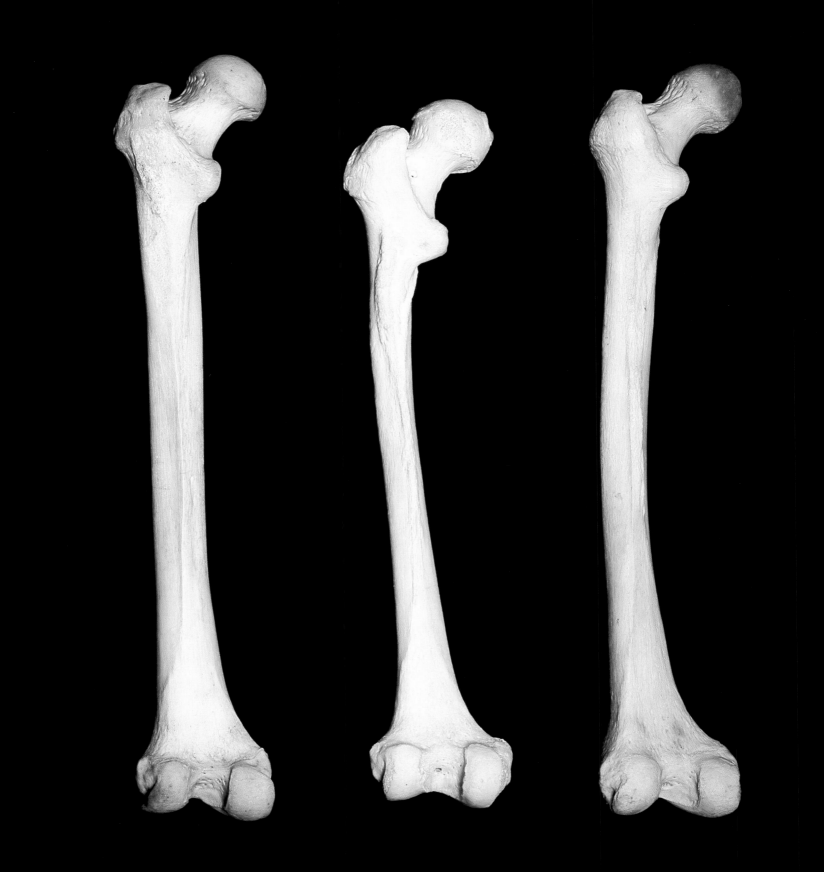

The archaeological investigation in the church at Spitalfields, mentioned in the previous chapter, gave an opportunity for checking methods of estimating age from skeletons. Ages at death were estimated by various methods, and compared with the actual ages given by the coffin plates. Many of the estimates were found to be five to ten years too high, presumably because people who have aged prematurely are likely to die young.

## RACE

As well as sex, stature and age, skeletons can provide evidence of race. The major racial groups can generally be identified from skulls, but the distinctions are not always clear cut, and mixed ancestry can cause confusion. The major groups are Negroid (originally from Africa south of the Sahara), Mongoloid (from northern and eastern Asia, together with Eskimos and American Indians) and Caucasoid (from Europe, the Arab countries and the Indian subcontinent). [Figure 17] Australian aboriginals and Polynesians (including Maoris) are smaller, distinctive groups. Negroid skulls tend to be relatively long and Mongoloid skulls relatively broad, compared to Caucasoid skulls [Plate 92]. Negroid skulls generally have wide nasal apertures, rather rectangular eye sockets and protruding jaws. Mongoloid skulls tend to have rounder eye sockets, less prominent brow ridges and broad zygomatic arches that give the face a characteristic appearance with high cheekbones. Most mongoloid skulls also have shovel-shaped upper incisor teeth, concave on the rear face. Differences like these can be expected to arise by chance, between populations that remain separated for thousands of generations.

Subgroups can be distinguished, within the major racial groups. For example, the tall, blond Teutonic people of northern Europe tend to have long (dolichocephalic) skulls, while the typically shorter, brown-haired Celtic people of central Europe have rounder (brachycephalic) skulls. [Plate 93]

## SKULL RECONSTRUCTIONS

Use of skulls to reconstruct the appearance of living people is a commonplace of archaeological television programs. It has been refined for forensic use, enabling the police to produce pictures that may be recognized by friends of the deceased. The reconstructions depend on knowing how thick the skin and muscle are likely to have been, on the living skull. The modern techniques known as computer-aided tomography and magnetic resonance imaging make it much easier than in the past to collect detailed, accurate data from living people. These methods record the three-dimensional structure both of the skeleton, and of the soft tissues around it. Reconstruction of faces has traditionally been done by spreading appropriate thicknesses of clay over a cast of the skull, but can now be done on a video screen. A

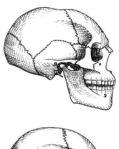
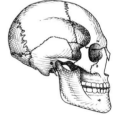
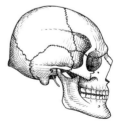

*Figure 17 Three skulls: Negroid (top), Mongoloid (middle) and Caucasoid (bottom)*

*Plate 86 (opposite) Individual differences can be seen between other bones, for example these left femurs, as well as between skulls. (Courtesy Maxilla & Mandible.)*

three-dimensional image of the skull is obtained by laser scanning and displayed on the screen. The soft tissues are added electronically, and the result can be viewed from any angle. Important limitations of the technique are that the skull gives no information about the colors of the skin, hair or eyes. If, however, the race of the skull can be determined, the range of possibilities will be limited. The skull also tells us nothing about hair styles or facial hair.

## DATING SKELETONS

A question that often arises both in criminal investigations and in archaeology is how long ago the owner of a skeleton died. The appearance of bones can be misleading because it depends so much on the environment. Flesh decays at different rates in different circumstances. Dried tissue may remain for many years on a skeleton in a vault, but if the body had been left exposed to the weather almost all traces of soft tissue would generally have disappeared within a year. After the flesh has gone, organic matter (mainly collagen) still remains within the bone. It decays gradually, leaving only the bone mineral (hydroxyapatite). The bone becomes lighter in weight, and more liable to crumble. The stage of decay can be assessed by chemical analysis. Collagen and other proteins are long chains of amino acids, which contain nitrogen. The nitrogen content of old bones gradually falls, halving within the first three or four centuries after death. Different amino acids disappear at different rates, the fastest in about 50 years and the slowest over many centuries. A cut surface of a fresh bone fluoresces in ultraviolet light. As the organic matter decays, the fluorescence is progressively lost, starting with the outer layers of the bone.

Murderers are unlikely to survive for more than 70 years or so after their crimes, so forensic scientists are not interested in precise dating of skeletons older than that. Archaeologists, however, want to know dates of death for skeletons that are centuries or even millennia old. Often the dates of skeletons are apparent from things found with them, making it unnecessary to seek the evidence of the bones themselves. The date of an earthenware pot may be estimated on stylistic grounds or measured by the technique of thermoluminescence. Wooden objects can be dated by the pattern of wide and narrow annual rings, reflecting different rates of growth due to variation of the weather from year to year. Coins may have dates on them. Often, however, direct evidence of the date of a skeleton is needed.

Several methods are used for measuring the ages of old bones. Carbon dating cannot cope with the relatively short times that interest forensic scientists, but has been immensely valuable for measuring the longer times that are important to archaeologists. Carbon contains a tiny proportion of the isotope carbon-14, which is formed from nitrogen atoms in the upper atmosphere by the action of cosmic rays. Carbon-14 is unstable and changes back to nitrogen at a constant

*Plate 87 (opposite) These forearm bones, from two men, are not identical. (Courtesy Division of Anthropology, American Museum of Natural History.)*

*Plate 88 (following pages) The man's skull (on the right) has a squarer jaw and more prominent brow ridges and chin than the woman's one (on the left). (Courtesy Division of Anthropology, American Museum of Natural History.)*

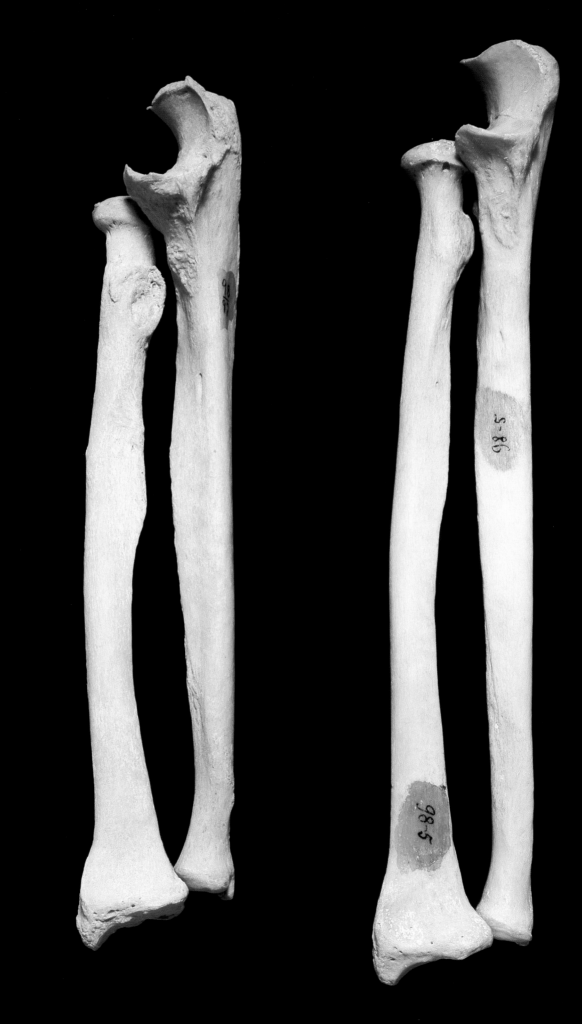

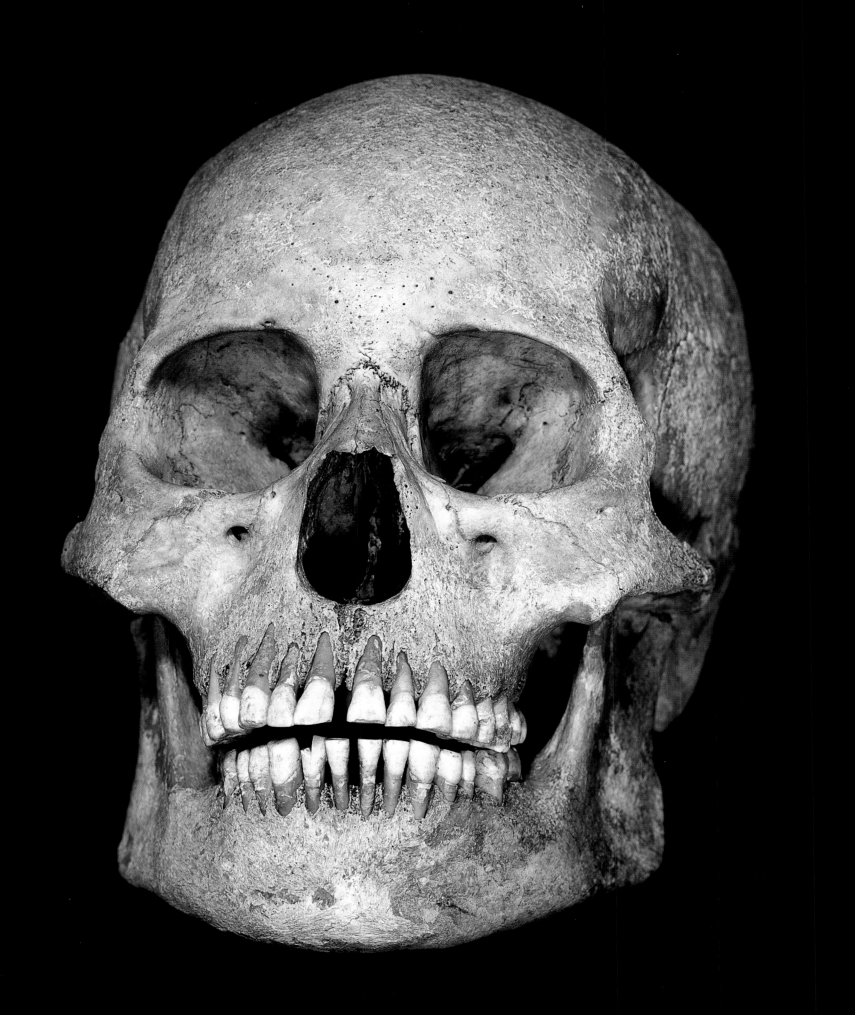

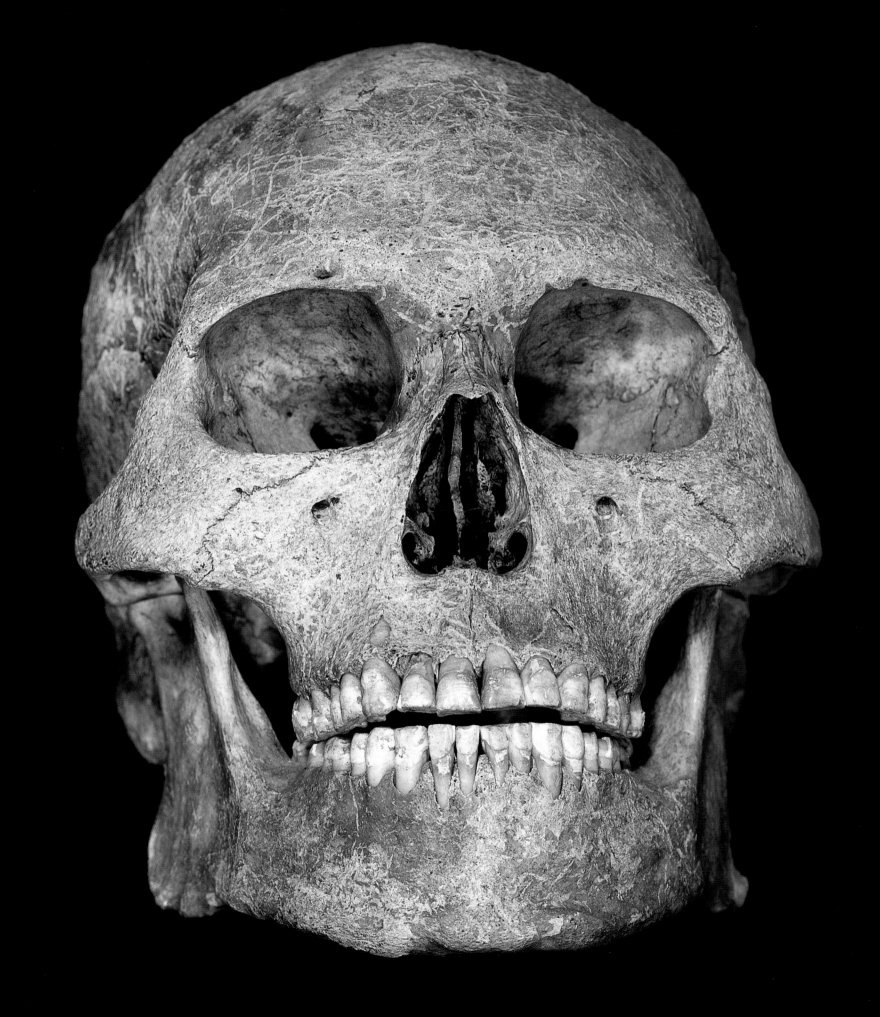

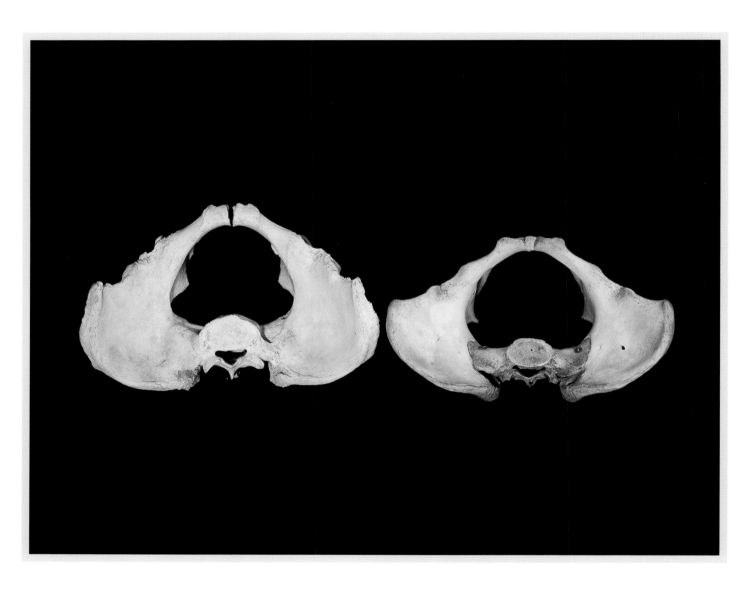

PLATE 89

*The woman's pelvis (on the right) is better shaped for*
*giving birth, than the man's pelvis (on the left).*
*The woman's birth passage is relatively larger than the*
*corresponding opening in the man. (Courtesy Division of*
*Anthropology, American Museum of Natural History.)*

rate, emitting radiation that can be detected by a Geiger counter or by accelerator mass spectrometry, which works on smaller samples. The proportion of carbon-14 in a sample of carbon falls to half its initial value in 5730 years, to one quarter in 11,460 years (twice 5730), to one eighth in 17,190 years, and so on. During their lives, living things exchange carbon with the atmosphere. Plants convert carbon dioxide from the atmosphere to foodstuffs. Animals and people eat foods that derive ultimately from photosynthesis, and get the energy they need by oxidizing these foods, producing carbon dioxide that is released back into the atmosphere. These processes keep the proportion of carbon-14 in the body constant. At death, however, this exchange ceases. The carbon trapped in bones gradually loses its carbon-14. The proportion remaining can be measured and the date of death calculated. There is always a range of uncertainty, which in different circumstances may be as little as a few decades or as much as a few centuries.

Carbon dating works well for up to 50,000 years after death. After that, there is too little carbon-14 to give reliable results. Other radioactive isotopes that decay more slowly can be used to measure the longer intervals of time that interest paleontologists. These measure the ages of the rocks in which the fossils are found, rather than of the fossils themselves. The most useful is potassium-40, which changes to argon-40, halving in concentration in 1300 million years. The argon, which is a gas, remains trapped in the rock unless the rock is melted, so any argon-40 found in a volcanic rock has been formed since the rock solidified. The method works only for volcanic rocks, not for the sedimentary rocks in which fossils are usually found. Sedimentary rocks, however, can often be dated by reference to volcanic ones formed at about the same time. In Chapter 7, we will examine fossil hominids who lived in East Africa a few million years ago. The sediments in which some of their bones were found are interleaved between layers of solidified lava from eruptions that occurred at intervals of a few hundred thousand years. Potassium-argon dating of the lava layers enabled paleontologists to establish approximate dates for the fossils.

Analysis of bones and teeth can provide evidence of diet as well as of age. In addition to ordinary carbon (carbon-12) and a tiny proportion of the unstable carbon-14, the carbon dioxide in the atmosphere contains about 1% of a third isotope, carbon-13. Plants converting carbon dioxide to foodstuffs by photosynthesis do not use the isotopes in the same proportion as in the atmosphere, but favor carbon-12. Different kinds of plants favor it to different degrees. The first stage of photosynthesis for most trees, shrubs and temperate grasses is the formation of a molecule containing three carbon atoms. Accordingly they are called C3 plants. Tropical grasses such as maize form a molecule with four carbon atoms in the first stage of photosynthesis, and are called C4 plants. C4 plants use more carbon-13 than C3 plants do. Marine plankton and seaweeds use still more carbon-13. Herbivorous animals have different proportions of carbon-13

*Plate 90 (following pages) The man's humerus (left) is longer and more robust than the woman's. (Courtesy Maxilla & Mandible.)*

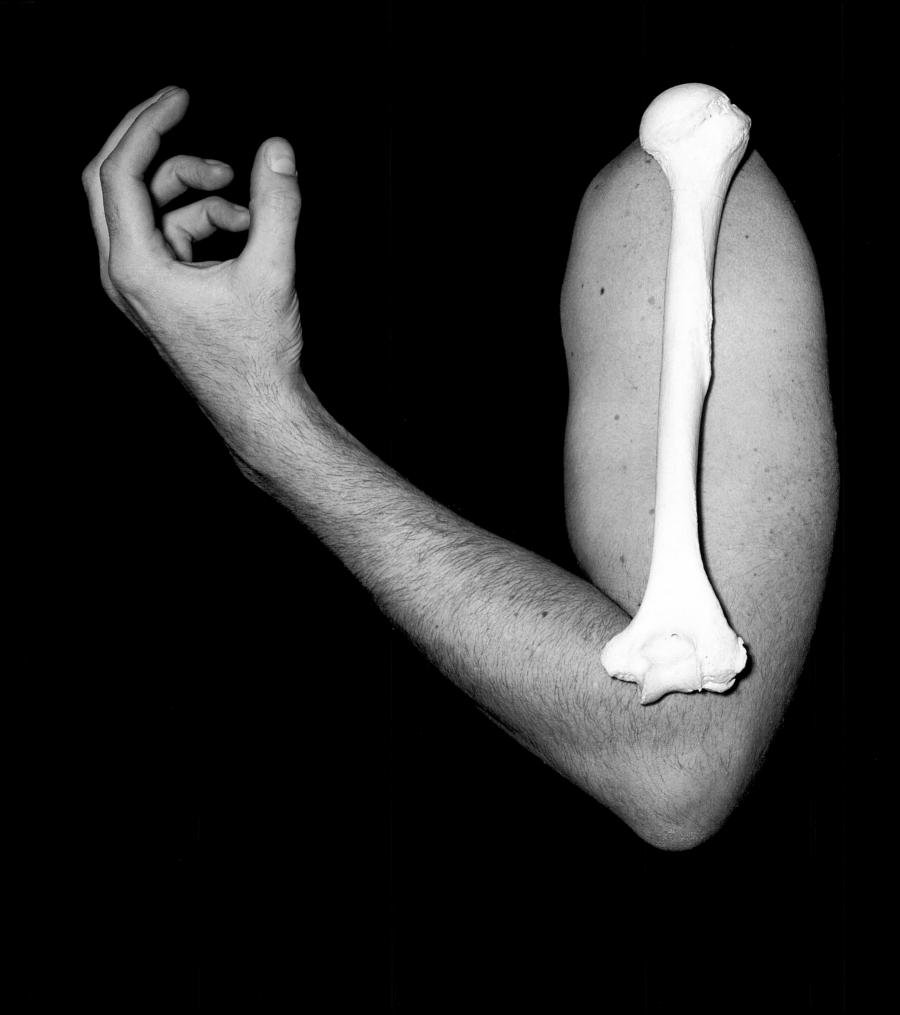

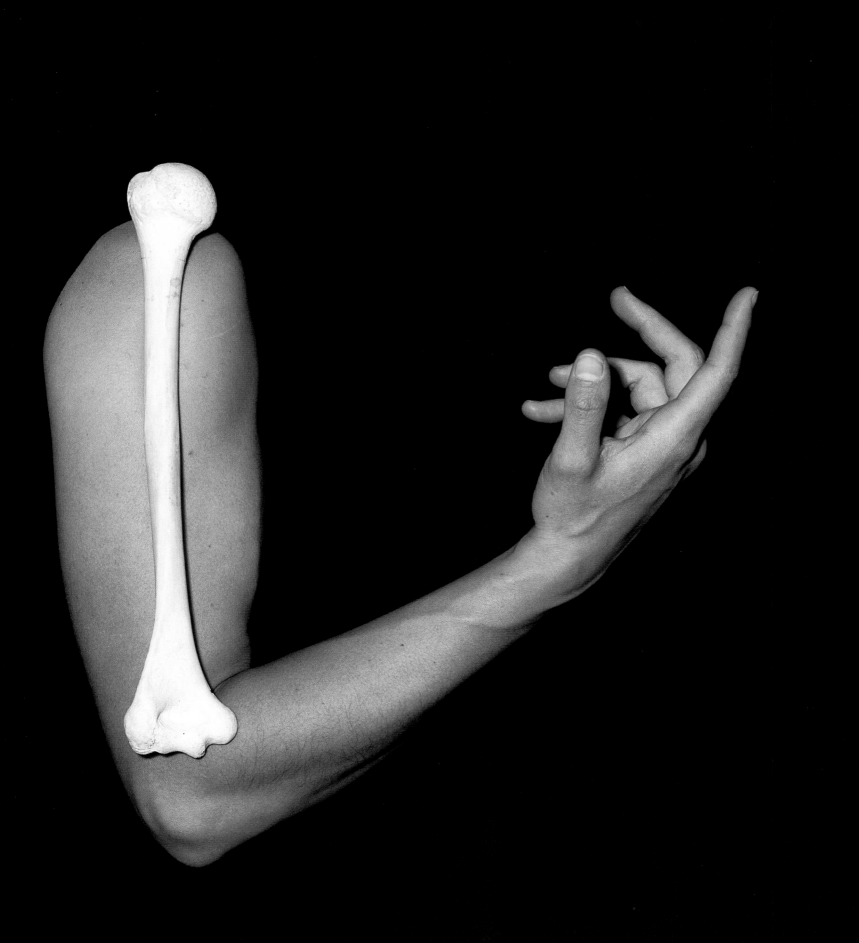

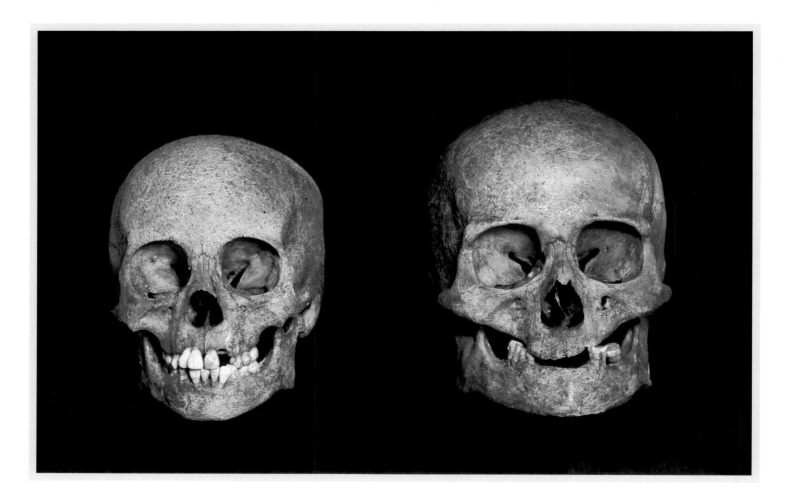

in their bodies, according to whether they have fed mainly on C3 plants, on C4 plants or on marine plants. Carnivores have the same proportions as the herbivores they prey on. Measurement of the ratio of the isotopes in human bones can show whether people fed mainly on food derived from C3 plants, from C4 plants or from the sea.

*Plate 91 Skulls of a young adult (left) and an aged one (right). (Courtesy Division of Anthropology, American Museum of Natural History.)*

For example, analysis of prehistoric skeletons from British Columbia shows that their food came mainly from the sea. Skeletons from Venezuela show a change from a predominantly C3 diet to a C4 one between about 800 BC and 400 AD. The analyses cannot identify the particular plant species involved but it appears from other evidence that the change was from crops such as cassava to maize. (Cassava is a plant that accumulates starch in its roots, and is the source of tapioca.) Investigation of bones from Ontario showed that the change from a C3 diet to maize occurred later there, between about 400 and 1650 AD.

Other analyses of bones can distinguish between vegetarian and meat diets. One method depends on isotopes of nitrogen in collagen, and another on the element

PLATE 92
*These skulls are from men of different racial groups:*
*(left) Mongoloid, (center) Negroid, and (right)*
*Caucasoid. (Courtesy Division of Anthropology,*
*American Museum of Natural History.)*

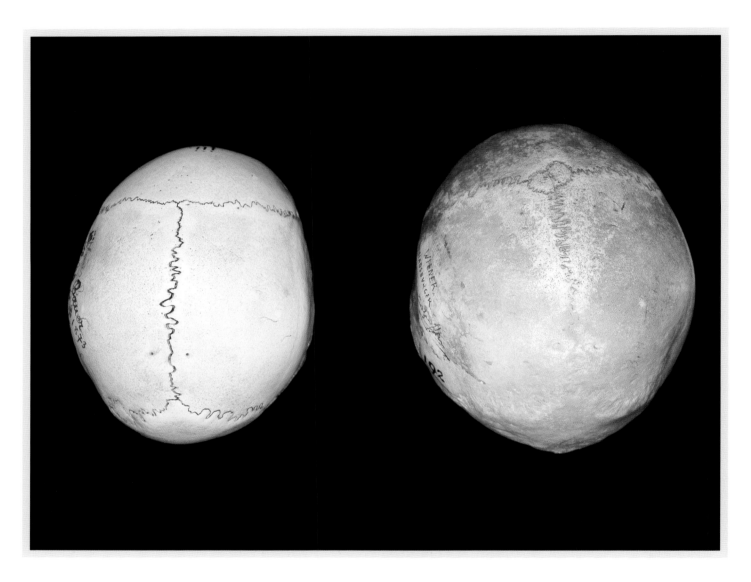

PLATE 93
*A dolichocephalic skull (left) and a brachycephalic one.*
*(Courtesy Division of Anthropology, American*
*Museum of Natural History.)*

strontium. Strontium is chemically similar to calcium, and plants accumulate the two elements indiscriminately. Animals, however, excrete strontium, allowing only a small proportion of the strontium in their food to remain in the body. As a result, vegetarians have more strontium in their bones than meat eaters. This technique was applied to Olmec skeletons of 750-500 BC in Mexico. It was shown that people buried with expensive jade ornaments had eaten a lot of meat, while others buried without grave goods had eaten very little. Then as now, the rich had a more luxurious diet than the poor.

This chapter has been concerned with the differences between people of our own species, *Homo sapiens*. In the next we go back in time to look at our ancestors.

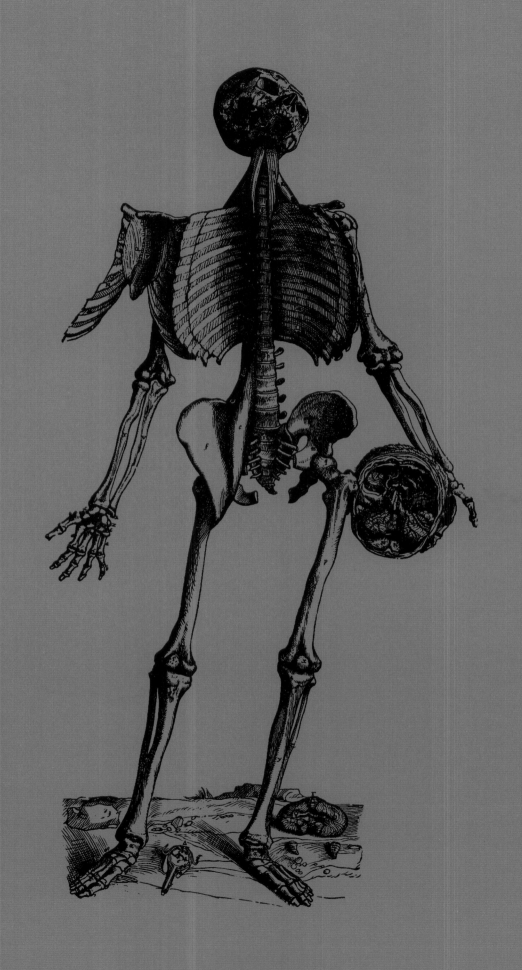

# 7

# THE EVOLUTION OF BONES

We humans are mammals, members of the great class of animals that also includes (among many others) dogs, cattle, elephants, bats, mice and whales. Zoologists divide the mammals into a series of orders that are believed to reflect evolutionary relationships. For example, dogs, lions, badgers, bears, hyenas and similar mammals are grouped together as the Order Carnivora. Mice, squirrels, beavers, guinea pigs and their relatives form the Order Rodentia. We are placed in the Order Primates with the lemurs, monkeys and apes.

## HUMAN ORIGINS

When Charles Darwin published *The Origin of Species* in 1859 it was clear that it implied that we had evolved from animals. Darwin avoided the sensitive topic of human origins until he wrote *The Descent of Man*, published in 1871. In *The Origin* he merely remarked that in the course of future research "much light will be thrown on the origin of man and his history," but the implication was obvious. Unhappily, the theory of evolution is still seen by some people as incompatible with their religious beliefs. The vast majority of professional biologists, however, are convinced by the evidence that has accumulated since Darwin's time. The theory of evolution is central to much of our work. In this chapter, I try to show how the human skeleton has been molded by evolution since the early days of the mammals.

The earliest fossils of mammals are about 200 million years old, about the same age as the earliest dinosaurs. While the dinosaurs thrived, mammals played only minor roles in the world's ecosystems. It was only after the extinction of the dinosaurs 65

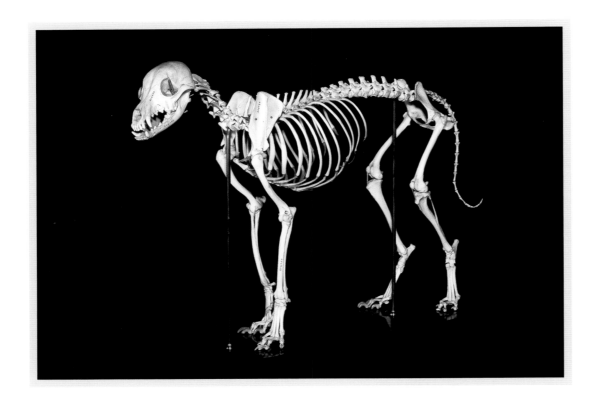

Plate 94 *The skeleton of a dog, an example of a mammal that is not a primate. (Courtesy Maxilla & Mandible.)*

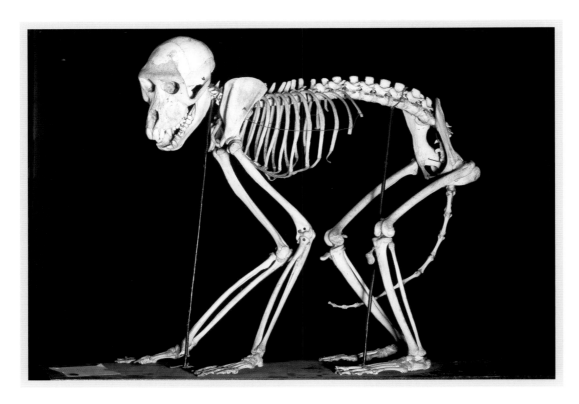

Plate 95 *The skeleton of a young female baboon, as an example of a monkey. (Courtesy Division of Anthropology, American Museum of Natural History.)*

million years ago that the mammals were able to take over their role as the most successful large land animals. Early mammals were small, in the size range of modern rodents. The earliest identifiable as primates appeared shortly before the extinction of the dinosaurs, but monkeys did not appear until about 35 million years ago and humans (as we will see) much later.

We can trace our ancestry from early mammals through monkey-like mammals and apes to early fossil humans and (eventually) ourselves. In this chapter we will look at some of the main stages in our ancestry. We will start by comparing the skeletons of a dog, a monkey and an ape. [Plates 94, 95 and 96] None of these is exactly like any of our direct ancestors, but the complete skeletons of the dog, monkey and ape give reasonably good impressions of stages in our ancestry. The common ancestor from which apes and humans evolved was more like an ape than a human. Further back, the common ancestor of monkeys, apes and humans was more like a monkey than an ape or a human. The sequence from dog through monkey and ape to human gives a reasonable first impression of our evolutionary history.

## DOGS AND MONKEYS

I have chosen the dog to start the sequence, as a familiar and in many respects typical mammal, although in other respects it is far from typical. Its teeth are specialized for a carnivorous diet, and its feet are specialized for fast running. A tree shrew or an opossum would have been more like our earliest mammal ancestors. Tree shrews and opossums are, however, smaller than most monkeys, and much smaller than the apes and early humans that we will discuss later in the chapter. We can avoid confusing complications at this stage by comparing animals of similar size. Later we will have to explore some of those complications.

At first glance, you may be struck more by the similarities between the dog and monkey skeletons than by the differences. There are differences in the skull that we will examine soon. The dog has shorter toes than the monkey, and stands with its metacarpal and metatarsal bones almost vertical, so that its wrists and heels are well clear of the ground. In contrast, the monkey stands with its metacarpals and metatarsals horizontal, so that the palms of the hands and the soles of the feet lie flat on the ground. In this respect, monkey's hands and feet are more like those of early mammals than the dog's ones are. The dog's stance, with the metacarpals and metatarsals vertical, seems to be an adaptation for fast running. It increases the effective length of the legs without adding to their weight. The one feature of the feet in which the dog is more primitive than the monkey is that its toes have claws, whereas monkeys have finger- and toe nails.

The chimpanzee skeleton [see Plate 96] is shown standing on its hind legs, which makes it look much more human-like than the dog and monkey. Chimpanzees,

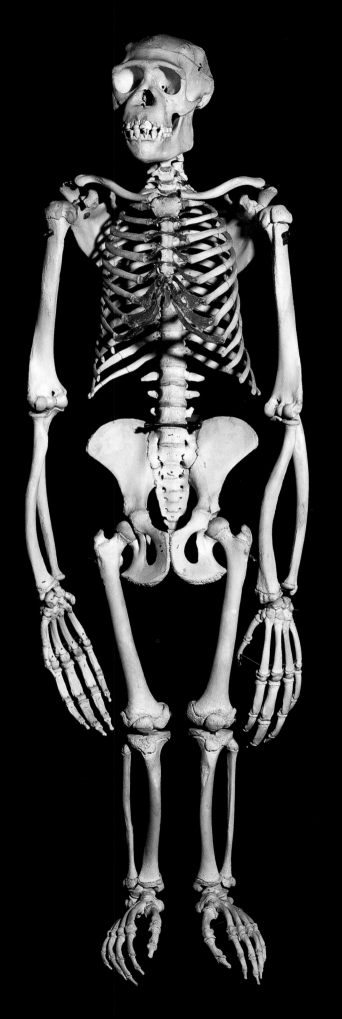

*Plate 96 The skeleton of a chimpanzee.*
*(Courtesy Division of Mammology,*
*American Museum of Natural History.)*

however, spend only a little of their time standing bipedally. In disputes, they get up on their hind legs to make themselves look more threatening. They occasionally stand bipedally to see over vegetation, or walk bipedally to leave their hands free to carry something, but they usually stand and walk on all fours. There is one major difference between their style of quadrupedal walking and that of monkeys. Monkeys walk with their fingers and the palms of their hands flat on the ground, but chimpanzees walk on their bent knuckles, with their wrists well off the ground.

A front view of a monkey skeleton [see Plate 97] shows that the rib cage is deep and narrow. Dogs and most other mammals have rib cages of similar shape, but humans [see Plate 65] and apes have relatively much broader chests. The scapulae are held against the sides of the narrow chests of dogs, monkeys and most other mammals, but lie flat against the broad backs of apes and humans. Consequently, the socket in the scapula, into which the head of the humerus fits, points down towards the ground in dogs and monkeys, but out to the sides of apes and humans. The difference has been related to different requirements for locomotion. A dog running on the ground keeps its feet below its body, with no need to reach out to the sides. The same is true for a monkey running along the tops of branches, which is how most monkeys usually move through trees. An ape clambering among branches, however, needs to reach out in all directions to take advantage of all available handholds. Scapulae that point sideways allow a greater range of movement at the shoulder.

Climbing style necessarily depends on an animal's size. Small monkeys, like squirrels, must often find themselves with only one branch within reach. Their only options are to run along the top of the branch, to swing below it or to jump to another branch. If the branch is horizontal, and thick compared to the size of its hands and feet, running along the top of the branch is easier than swinging below it. If most of the branches are near-vertical, jumping from branch to branch may be the best option. Bushbabies (in Africa) and lemurs (in Madagascar) live in forests where the branches are preponderantly vertical, and generally travel by jumping from branch to branch. Squirrels and the smaller species of monkey generally run along the tops of branches, jumping only to get from one tree to the next. The great apes (chimpanzees, gorillas and orangutans) are larger than most monkeys and often move among branches that would be inconveniently small to run along. For them, clambering is often the best way to move through trees.

The gibbons of south east Asia (the lesser apes) and the spider monkeys of South America are much smaller than the great apes. Most of them weigh 7 kilograms or less, but they live and feed among the small branches high up in the canopy that are too small for them to run along. To travel, they swing by their long arms from one handhold to the next. Like the great apes, they have evolved broad chests with the scapulae on their backs rather than their sides. The sockets on the scapulae, where the

*Plate 97 (following pages) This baboon's chest is deep and relatively narrow, with the scapulae on the sides rather than the back. (Courtesy Division of Anthropology, American Museum of Natural History.)*

heads of the humeri articulate, point upwards and outwards, giving the shoulders the mobility that they need for swinging from branch to branch.

## CONSTRAINS OF ANCESTRY

Many scientists would criticize the line of argument of the previous few paragraphs that interpret the chests and shoulders of monkeys and apes as adaptations to different styles of climbing. Their objection was most famously expressed in a paper published in 1979, by Stephen Jay Gould and Richard Lewontin. These authors criticized the tendency of some biologists to interpret every feature of an animal's structure as a beautiful adaptation to some aspect or other of its way of life. They condemned "adaptationist" biologists for being so blind to other kinds of explanation that as soon as one adaptationist explanation is discredited they will substitute another: any plausible story will do. Gould and Lewontin emphasized a point that I have made repeatedly in this book, that the structures of animals are constrained by their ancestry. An ape might be better able to travel through forests if it had feathered wings like a bird, or eight legs like a spider, but its range of evolutionary opportunities is limited by its present structure. It is generally agreed that Gould and Lewontin's paper made a valid and important point, but it is I think widely considered that they overstated their case. The concept of adaptation is a valuable aid to understanding the structure of animals, and our own evolution.

## ADAPTATIONS

That digression was prompted by the suggestion that the broad chests and outward pointing scapulae of apes may be adaptations for clambering and swinging through trees. In repeating the suggestion I am not claiming that that evolution of these features was driven by natural selection for more effective locomotion through trees. Nor am I claiming that ape's chests and shoulders have the best possible structure for their style of movement. I am simply saying that they seem better suited to it than typical monkey-like chests and shoulders would be.

We humans, like our ape ancestors, have broad chests, with the scapulae on our backs rather than our sides. We presumably got that arrangement as a legacy from our ape-like ancestors. It does, however, have the advantage of making it easy for us to reach out with our arms in (almost) all directions. There is, however, a marked difference between our broad chests and those of apes. In apes the rib cage tapers towards the top, but in humans it is more barrel-shaped.

Chimpanzees can walk quite well on their hind legs, but they seem unable to run bipedally. To go fast, they get down on all fours and gallop. The reason that they cannot run bipedally may be that their legs are not strong enough. When we walk,

*Plate 98 (opposite) The pelvis of a chimpanzee. (Courtesy Division of Anthropology, American Museum of Natural History.)*

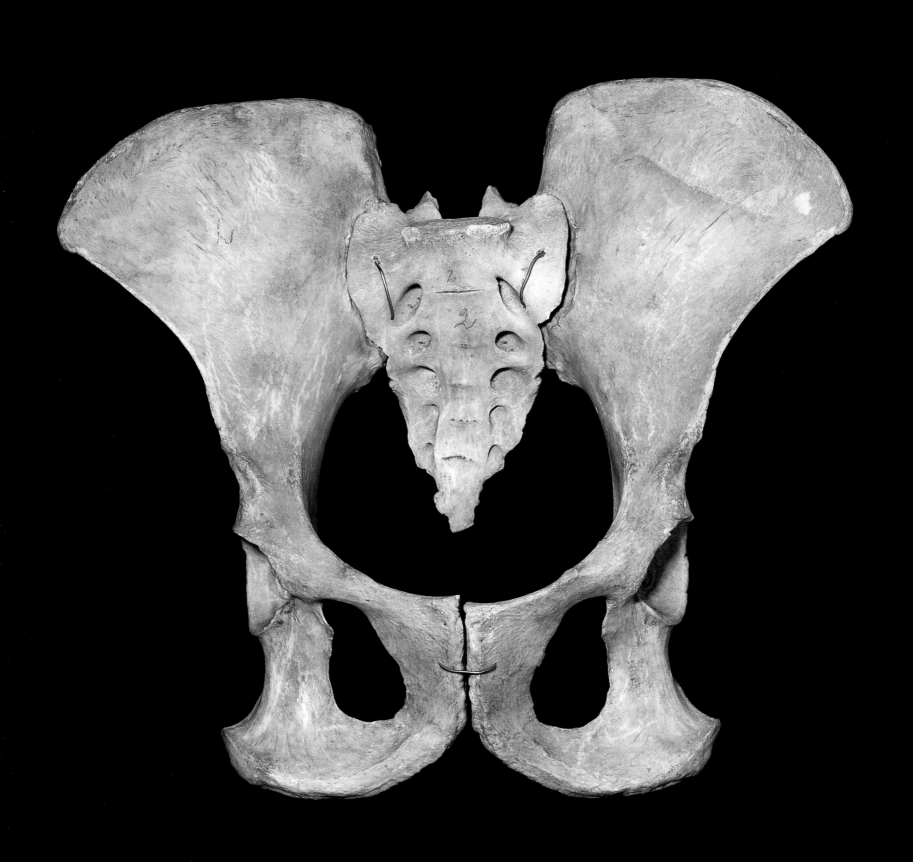

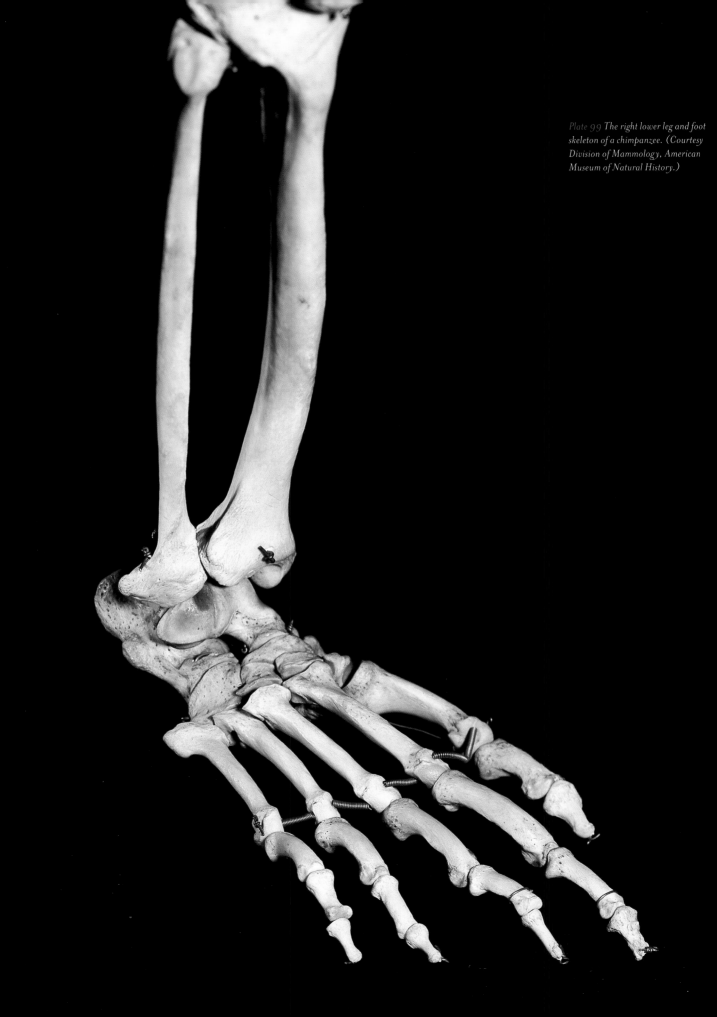

Plate 99 *The right lower leg and foot skeleton of a chimpanzee. (Courtesy Division of Mammology, American Museum of Natural History.)*

each foot is on the ground for a little more than half the time. When we run, each foot is on the ground for only one quarter to one third of the time. If the feet are off the ground for more of the time, they must exert larger forces while they are on the ground. Experiments in which people walk or run over instrumented panels in the floor show that the peak forces exerted by a foot are 1.0 times body weight in walking, about 2.7 times body weight in running at marathon speed and more in sprinting. The faster we go, the stronger our leg muscles need to be.

My former colleague Sue Thorpe dissected chimpanzees that had died in zoos, weighing and measuring their muscles and calculating the forces they could have exerted in the living animal. She found that their leg muscles were considerably weaker than we would expect to find in a human of the same body weight. The arm muscles, however, were much stronger than expected for a human. Accordingly, chimpanzees are less good than we are at bipedal running, but they are better at swinging through the trees.

## BIPEDS VS. QUADRUPEDS

The problem of leg muscle strength is made worse by a difference between chimpanzee and human styles of walking. [Figure 18] We walk with the trunk vertical, and we keep each leg straight while its foot is on the ground. Chimpanzees walk bipedally with their knees bent and the back sloping forward. Try walking around for a while like a chimpanzee. You will find that the muscles on the fronts of your thighs get tired. These are the muscles that extend the knees. When the knee is straight, the leg bones function as a pillar and support your weight with little need of effort from the muscles. When the knee is bent, however, the extensor muscles must be active to prevent your knees from buckling under your weight. They have to exert bigger forces to support your weight when your knees are bent, than when they are straight.

Energy is needed to develop the increased muscle forces that are needed for bent-legged walking. Measurements of oxygen consumption show that chimpanzees use more energy when walking than humans of equal weight would do. In an experiment in which people walked normally and with bent knees, it was found they used more energy when imitating chimpanzees.

It may seem that the chimpanzee has an easy solution to the problem. It would need less energy and less force from its knee muscles if it walked and ran like us, on straighter legs. That does not seem to be a real option, because the leg movements that the chimpanzee can make are limited by the structure of its pelvis [Plate 98], which has the same shape as in quadrupedal mammals such as dogs and monkeys. Pelvis shape limits the range of hip movements, because the muscles that move the hip have their upper attachments on the pelvis. The chimpanzee pelvis is well suited to quadrupedal

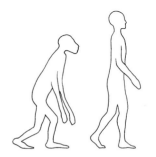

*Figure 18 Bipedal walking postures of a chimpanzee and a human.*

Plate 100 *Skulls of a male baboon (left) and a dog, seen from above. (Courtesy Division of Anthropology, American Museum of Natural History.)*

walking but not, as we will see, to the human style of bipedalism. Human pelvic girdles have a very different shape. [see Plate 67]

As a quadruped walks with its back more or less horizontal, its legs swing forward and back, making a mean angle with the back of about 90°. In a biped walking with the back erect, the mean angle of the legs to the back is about 180°. The chimpanzee pelvis provides suitable attachment points for muscles to move the hip through the range of angles required in quadrupedal walking. It does not provide well-placed attachments for muscles to work the joint in the very different range of angles required for human-like walking. By keeping its back sloping and its knees bent, the chimpanzee manages to walk bipedally without having to extend its hip joint to 180°.

In humans, the ilium (the blade of bone that connects to the sacrum) is broader and shorter than in other mammals. The ischia (the bones in our bottom, which we sit on) are bent back relative to the upper parts of the girdle so as to project well behind the hip joints. They provide muscle attachment points well behind the joints, enabling our legs to operate at angles around 180°.

The feet of monkeys and apes and of our monkey-like and ape-like ancestors look very different from human feet. [Plate 99] The main difference is that the big toe is as separate from the rest of the foot as the human thumb is from the palm of the hand. Consequently, it can be used like a thumb to grasp branches or other objects. Monkeys and apes can use their feet as additional hands. Georges Cuvier, the great French zoologist, recognized this in the classification of animals that he published in 1817. He grouped the monkeys and apes together in the Order Quadrumana (which means "four hands"). He put humans in a different Order, the Bimana (two hands). Another difference between the feet of monkeys and apes, and human feet, is that monkey and ape feet are flat, with no arch.

Human feet do not work well as hands. They are little use for anything except walking on. Our hands, however, are just a little better than ape hands for manipulating things. For delicate, controlled movements we use precision grips, holding objects between thumb and index finger. For example, we hold pens, scalpels and keys in that way. Our hands are better built than ape hands for precision grips, in two ways. First, we have longer thumbs, better able to reach the tip of the index finger. Second, the structure of the joints of the hand enables us (unlike apes) to rotate the thumb as it approaches the index finger, so that they meet fingerprint to fingerprint.

## MONKEY VS. DOG SKULL

Now compare the skulls of a monkey and a dog (as a more typical mammal). [Plate 100] The monkey skull has more room for its bigger brain. Monkey brains are about twice as heavy as the brains of most other mammals of the same body weight. Much of the difference is in the cerebral hemispheres, whose importance was explained in Chapter 2.

Another difference between monkey and dog is more obvious in front view. [Plate 101] The dog's eyes face out to either side of the head, giving the animal an extensive view to the sides. The monkey's eyes face directly forward, like human eyes: notice how the eye sockets stare at you, out of the page. Consequently, monkeys and people have more restricted fields of view than dogs. They are less able to see things to the sides and behind them. There is, however, much more overlap between the fields of view of the left and right eyes. The overlap gives the stereoscopic effect that is useful for judging distance. The ability to judge distance accurately is particularly important for monkeys and other animals which jump between branches, high up in trees.

A doglike snout on a monkey's face would partly obscure its forward view. Most monkeys have very small snouts. Baboons have long snouts, but they are set low on the face, quite well out of the way. [Plate 102] The monkey and dog skulls seen from below [Plate 103] show a difference related to the ways in which the animals hold their heads. In dogs as in most other mammals, the neck attaches to the back of the skull. Accordingly, the hole from which the spinal cord emerges (the foramen magnum) is at the back of the skull. Monkey necks attach to the underside of the skull and are held more vertical. The foramen magnum in monkeys, as in people, is on the underside of the skull rather than the back.

Though some of our pre-monkey ancestors had rather dog-like skulls, none had particularly dog-like teeth. Dogs have teeth like the wolves from which they are descended, suitable for killing prey and eating flesh. Other carnivores have similar teeth, but these teeth are not typical of mammals in general.

The grinding teeth of cattle, the gnawing teeth of rodents and the simple spikes in the jaws of dolphins (good for gripping slippery fishes and squids) are very different kinds of teeth suitable for different diets. Monkeys have teeth very similar to human teeth, with one major difference: the canines may be long fangs, especially in males. Many monkeys feed mainly on fruit, some feed principally on leaves and some eat a lot of insects. The relatively simple teeth of monkeys, which we have inherited, seem to have evolved from the more elaborate teeth of lemurlike ancestors.

## BRAIN SIZE

For the rest of this chapter we will be concerned largely with skulls and with the brains they contained. We will be making comparisons between animals of different sizes, and we will need to be clear about the relationship between brain size and body

Plate 101 *The dog (left) and baboon (right) skull seen from in front. (Courtesy Division of Anthropology, American Museum of Natural History.)*

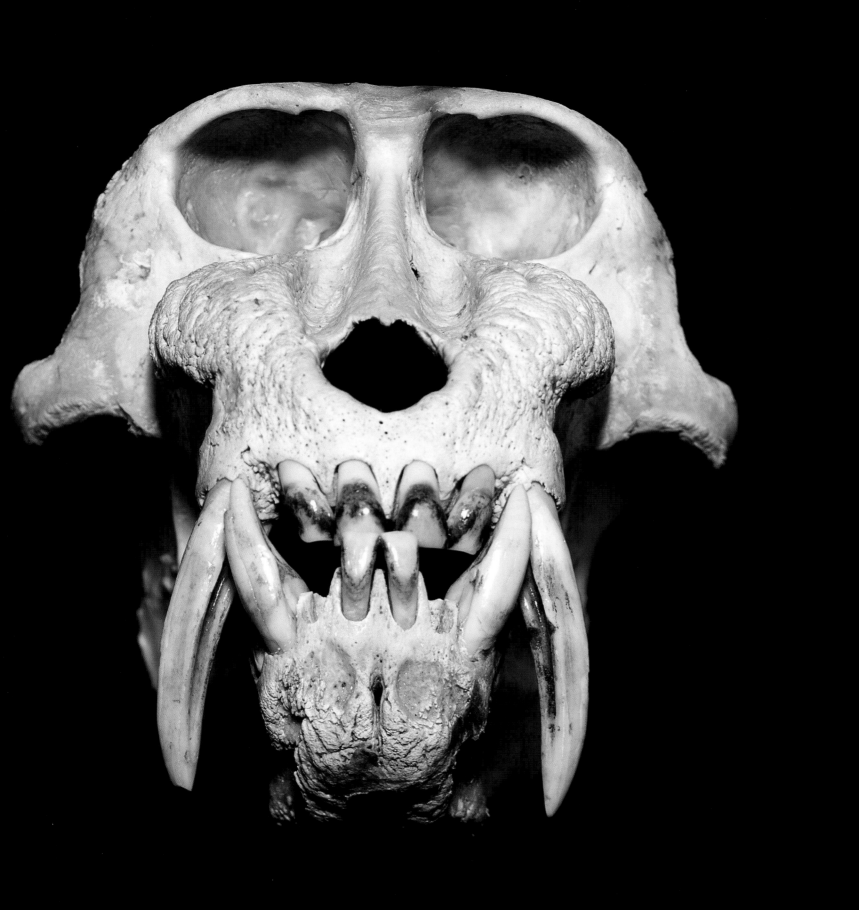

PLATE 102
*Skulls of a male baboon and a (smaller) colobus
monkey. (Courtesy Division of Anthropology,
American Museum of Natural History.)*

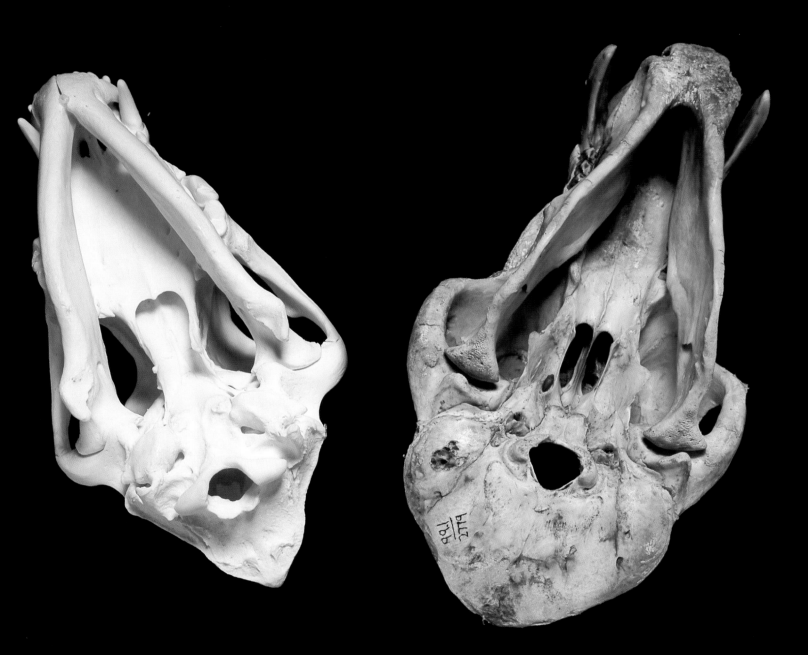

*Plate 103 The dog (left) and baboon (right) skulls, from below. (Courtesy Division of Anthropology, American Museum of Natural History.)*

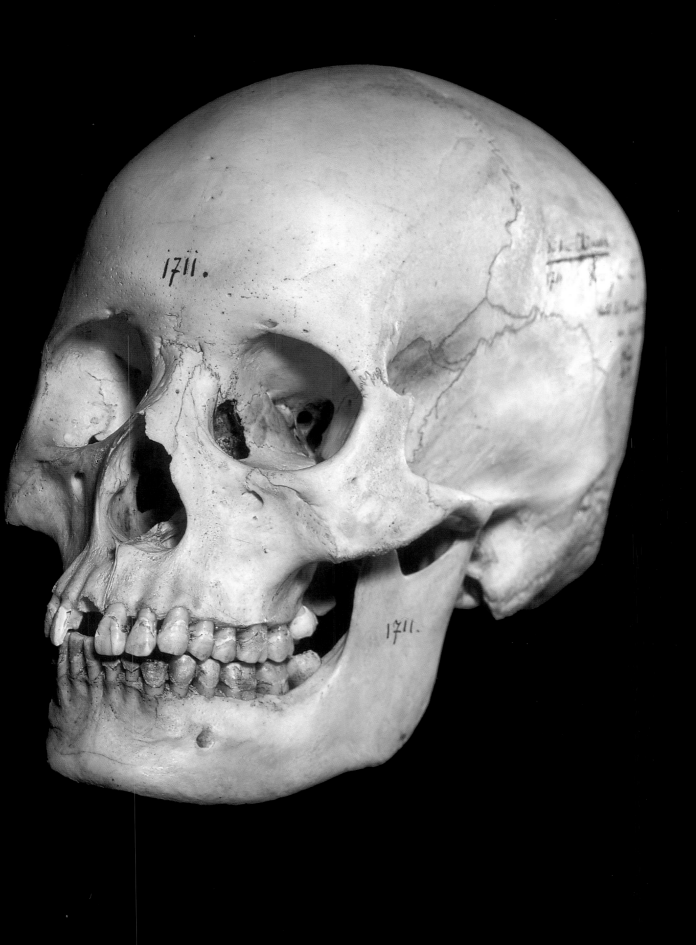

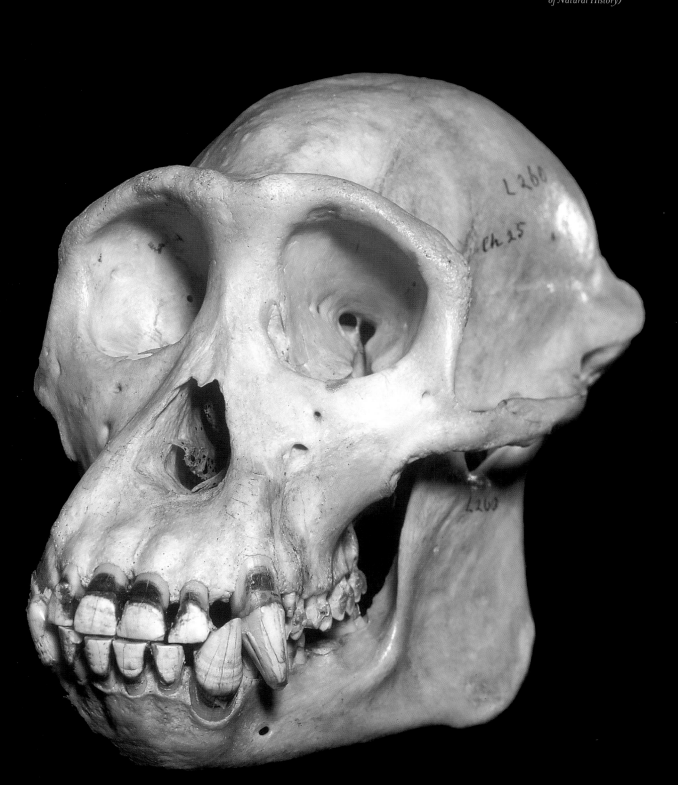

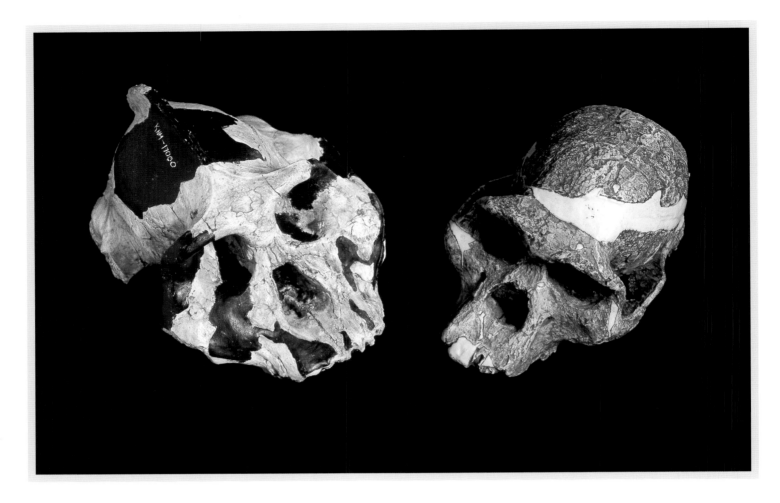

size. Plate 102 shows the skulls of two monkeys, of different sizes. The small one has a rounded braincase that makes it look like a miniature human skull. In contrast, the braincase of the big monkey looks shrunken. The swollen braincase of the small monkey provides ample area for the attachment of jaw muscles, but the braincase of the big one has been extended by bony crests for jaw muscle attachment. When the monkeys were alive, the brains of both would have filled their braincases with little space to spare. The brain of the small monkey was a bigger fraction of body weight than the brain of the large one. Indeed, the brains of some even smaller monkeys are about the same fraction of body weight as a human brain. That is why small monkey skulls look so human. There is nothing in the behavior of monkeys to suggest that small ones are any more intelligent than large ones. They certainly have nothing approaching human mental ability. How can we explain the relative size of their brains?

You might suppose that animals of different sizes but similar mental ability would have brains in proportion to their body weight. You might expect a 3 kg cat to have a brain 1000 times as heavy as a 3 gram shrew, and a 3 ton elephant to have a brain 1000

*Plate 105 Restored casts of skulls of* Paranthropus aethiopicus (*left*) *and* Australopithecus africanus. *The dark parts of the* Paranthropus *skull and the pale parts of the* Australopithecus *are restorations of parts missing from the fossils.* (*Courtesy Division of Anthropology, American Museum of Natural History.*)

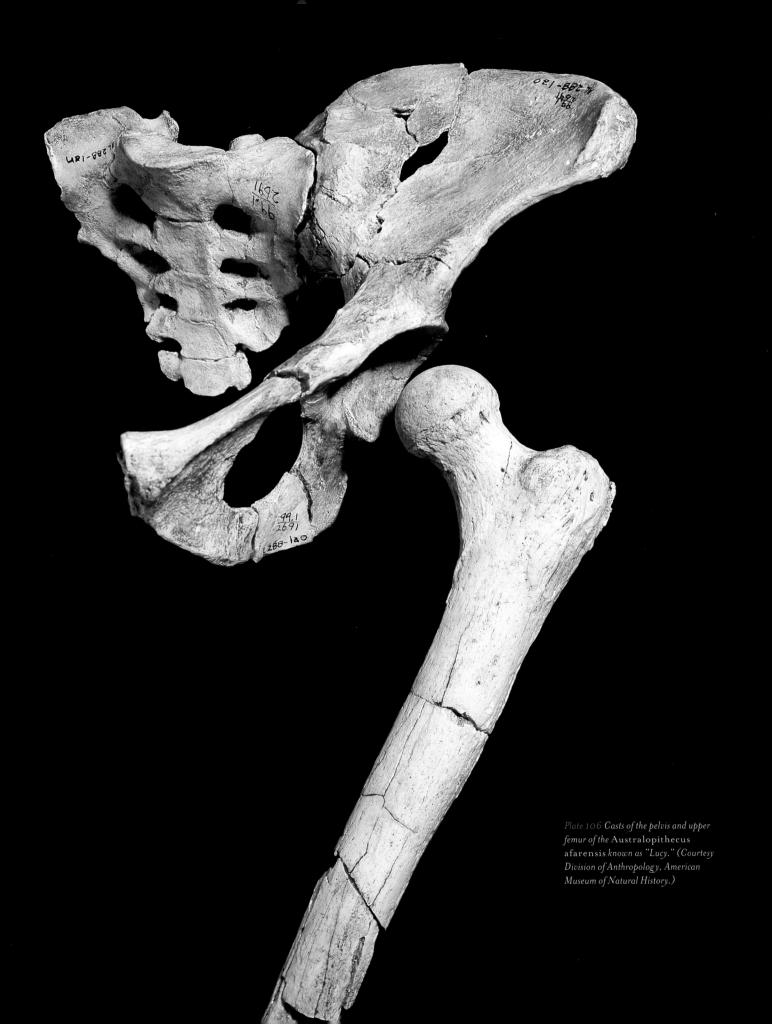

Plate 106 *Casts of the pelvis and upper femur of the* Australopithecus afarensis *known as "Lucy." (Courtesy Division of Anthropology, American Museum of Natural History.)*

# FIGURE 19: TRACING HUMAN EVOLUTION

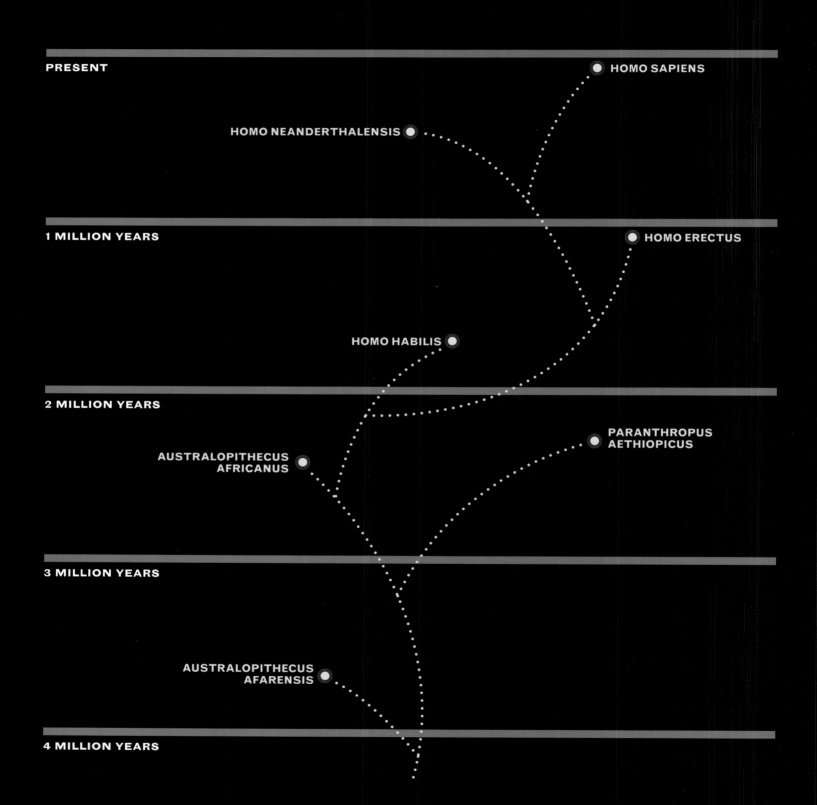

times as heavy as the cat. A different line of argument suggests, however, that a shrew-sized brain should be adequate for an elephant. The complexity of the operations that a brain can perform depends on the number of brain cells and on how they are interconnected. Cells are roughly the same size in animals of all sizes, so the same number of cells can be fitted into a gram of shrew brain and a gram of elephant brain. So why have elephants evolved larger brains than shrews?

A big brain, with more cells, can process more information from the sense organs, and control more complex behavior. So it is good to have a big brain. On the other hand, gram for gram, brain uses metabolic energy faster than most other tissues. So big brains are expensive to run. Natural selection presumably favors a compromise between performance and running costs. The energy cost of an extra gram of brain tissue is less significant for a large animal than for a small one, so we can expect larger animals to have bigger brains than smaller relatives.

On the other hand, larger bodies are no more difficult to control than small ones, so we should not be surprised that the brains of larger animals are a smaller proportion of body weight. I do not know of any plausible theory that would explain the relationship between brain size and body size more precisely than that. When animals and their brains have been weighed, however, a simple rule has been found to hold very generally. It applies equally well to comparisons between fishes of different sizes, or between non-primate mammals, or between monkeys. An animal that is ten times as heavy as an otherwise similar animal generally has a brain about 4.5 times as heavy. An animal that is 100 times as heavy has a brain about 20 $(4.5^2)$ times as big, and one that is 1000 times as heavy has a brain about 90 $(4.5^3)$ times as big. Another difference between the two monkeys in Plate 102 is that the bigger skull has relatively smaller eye sockets. This can be explained in the same sort of way as the difference in relative brain size. Bigger eyes give more acute vision, but there is no need for big animals to have eyes of the same relative size as small ones.

Plate 104 shows the skull of a chimpanzee. The size of the braincase shows that the brain was large, not as large in proportion to body weight as the brain of a small monkey, but about twice as big as would be expected for a monkey of chimpanzee size. Adult chimpanzees generally weigh 30-40 kg, and their brains have volumes of about 0.4 liters. By the rule I have just given, a man-sized (70 kg) chimp would be expected to have a 0.6 liter brain. Men's brains are around 1.4 liters.

## HUMAN FAMILY TREE

Chimpanzees are our closest living relatives. This has been shown by several lines of evidence, most convincingly by analysis of DNA. Gorillas are less closely related to us, and orangutans and gibbons still less closely. All the species that were more closely related to us than chimpanzees are extinct. For that reason, the remainder of this chapter is about fossils rather than the skeletons of living animals. Figure 19 shows how the fossils that we will look at may be related to each other and to ourselves. Some of the details of this family tree are controversial.

Scientists have given our own species the name *Homo sapiens*. *Homo* is the Latin for "man", and is the name of the genus (group of species) to which we belong. *Sapiens* (another Latin word) can be translated "rational". *Homo sapiens* is "rational Man". Other species similarly have two-word names, one word for the genus and one for the particular species. Evolution from ape-like ancestors to the genus *Homo* apparently occurred in Africa, and strenuous efforts there have uncovered a remarkable range of fossils that are close to our own ancestry.

## EXTINCT HUMANS

*Australopithecus afarensis* lived 3-4 million years ago, and its close relative *Australopithecus africanus* a little later. Their fossils have been found at various sites along the eastern side of Africa, from Ethiopia to South Africa. They were smaller than humans, 1.0 to 1.5 meters tall as adults. The small ones are believed to have been female and to have weighed about the same as a female chimpanzee. The large, presumably male ones seem to have been twice as heavy. Similarly, male gorillas are about twice as heavy as females.

Skulls of *Australopithecus* [Plate 105, right] do not look very different from chimpanzee skulls. The jaws project well forward in front of the eyes, as in apes. The volume of the braincase is only about 0.45 liters, which is within the range found in chimpanzees. The most obvious differences from chimpanzees are the less prominent brow ridges and the smaller canine teeth, both features that approach the human condition.

## "LUCY"

"Lucy" is one of the most famous of all fossils. She is an *Australopithecus afarensis* skeleton from Ethiopia, apparently a young female. About 40% of her bones have been found, far more than for any other *Australopithecus*. The most remarkable of these bones is surely the pelvis. [Plate 106] With its short, broad ilium it is very much more like a human pelvis than a chimpanzee one. The ilia, however, sloped out to either side of the body, giving Lucy very wide hips. Her ribs are all incomplete, but their curvatures seem to show that the rib cage tapered towards the top, as in apes. With broad hips and narrow shoulders, Lucy must have been pear-shaped.

One humerus and one femur of Lucy are complete. The humerus is about 0.8 times as long as the femur. In chimpanzees the humerus is relatively longer, and in humans it is shorter. Apes have long arms and short legs; humans have short arms and long legs;

*Plate 107 (opposite) Cast of a skull of Homo habilis. (Courtesy Division of Anthropology, American Museum of Natural History.)*

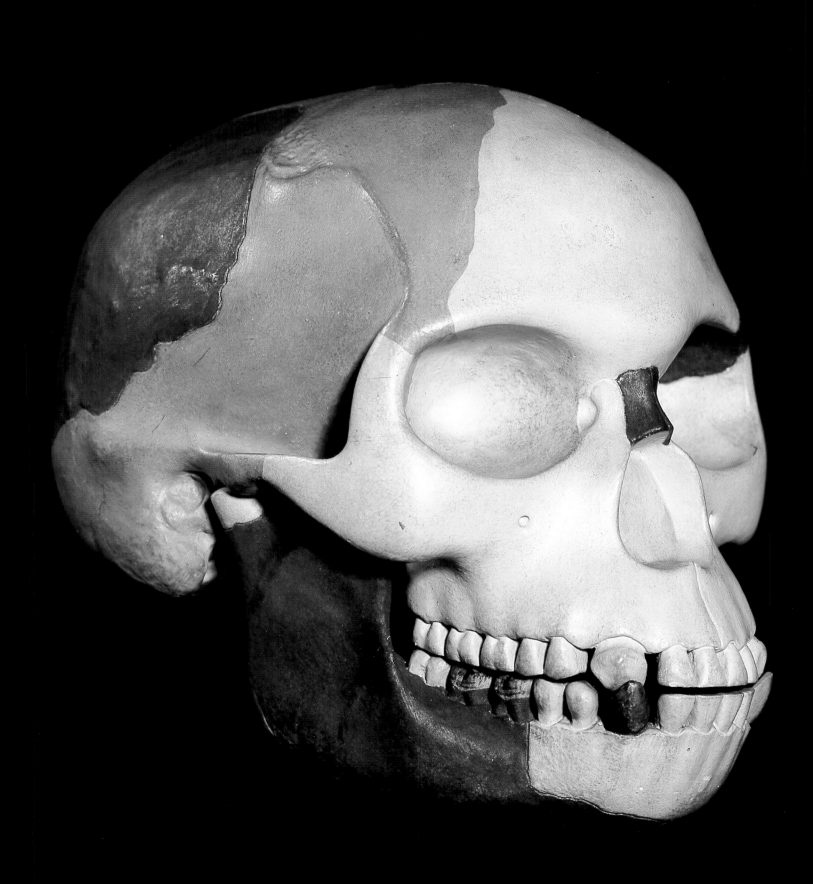

and Lucy was intermediate. Lucy's knees are more human-like than ape-like. We saw in Chapter 6 that our knock-kneed stance requires the lower end of the femur to be set slightly askew. This is not the case in apes, which stand with their two legs parallel to each other, with the hips, knees and feet all about the same distance apart. The shape of the lower end of her femur shows that the axis of Lucy's knee was set at an oblique angle, as in humans.

With her long legs and human-like pelvis, Lucy seems better suited to bipedal walking than to walking on all fours. Remarkably, there are footprints that seem to confirm that *Australopithecus afarensis* was indeed bipedal. They are at Laetoli, in Tanzania. They seem to have been made 3.6 million years ago, by *Australopithecus* walking over damp volcanic ash. The ash set like cement, and the footprints survived. There are prints of two individuals walking side by side over a distance of 24 meters. It has been estimated from the sizes of the prints that one (presumably male) was about 1.4 meters tall and the other (possibly female) 1.2 meters. Their speed can be estimated because people increase the length of their strides in a predictable way, as they increase speed. The estimate, which takes account of their being much less tall than adult humans, is about 0.7 meters per second (1.6 mph). This is very slow by human standards. The *Australopithecus* were plainly not in a hurry, on this particular occasion.

The Laetoli footprints are not very sharp, making it difficult to reconstruct the shape of the foot. The big toe, however, does seem to have pointed forward, not out at an angle as in apes. Hardly anything remains of Lucy's feet, but there is a group of four foot bones from another *Australopithecus* that seems to show that the big toe was well separated from the rest, as in modern apes. With her human-like pelvis and knees, Lucy probably walked more like a human than an ape, despite her ape-like feet.

## PARANTHROPUS

Three species of *Paranthropus*, close relatives of *Australopithecus*, lived in Africa at various times between one and three million years ago. *Paranthropus aethiopicus* lived at the same time as *Australopithecus africanus*, with which it is compared in Plate 105. They were about the same size, but *Paranthropus* was uglier, with a bigger lower jaw, bigger molar teeth and more pronounced brow ridges. Its brain was about the same size as in *Australopithecus*, but its skull was more heavily built and had a bony crest running along the top to give attachment to larger jaw muscles. *Paranthropus* must have had a very powerful bite. It may have eaten food that had to be cracked open. West African chimpanzees crack nuts open by hitting them with stones, but perhaps *Paranthropus* cracked nuts with its teeth. Many *Paranthropus* skulls have been found in a cave at Swartkrans in South Africa, among bones of antelopes and other mammals. The cave is connected to the ground above by a vertical shaft, and the bones were probably washed into it by water from

*Plate 108 (opposite) A restored cast of the forged Piltdown skull. The dark brown parts represent the original specimen; the light brown parts have been restored as mirror images of bone present on the other side of the head; and the pale parts were conjectural. (Courtesy Division of Anthropology, American Museum of Natural History.)*

rainstorms. At first it was assumed that the antelopes were the *Paranthropus'* prey, but it now seems that both the antelopes and the hominids are the remains of prey eaten by carnivores. One of the *Paranthropus* skulls has two punctures the right size and the right distance apart to have been made by the canine teeth of a leopard. Leopards taking antelope carcasses to a safe feeding place often grab hold of them by the head, and it seems likely that the unfortunate *Paranthropus* got the same treatment.

## HOMO HABILIS & HOMO ERECTUS

The extinct species that are closest to the human condition are placed with *Homo sapiens* in the genus *Homo*. *Homo habilis* [Plate 107] lived in East Africa 1.5-2 million years ago. The name means "handy man", and was chosen because stones that seem to have been deliberately chipped to form tools have been found in the same deposits. Also in these deposits there were antelope, zebra and giraffe bones with characteristic scratch marks on them. In experiments with fresh modern bones, stone tools made similar scratches. The obvious conclusion is that the tools were used for butchery.

Chimpanzees select suitable stones for cracking nuts, but do nothing to improve their shape. The distinction between chimpanzees as tool users and humans as toolmakers is not so sharp, if sticks as well as stones are regarded as tools. Chimpanzees dip twigs into ant and termite nests, and eat the insects that cling to the twigs. They prepare the twigs by stripping off the leaves. If the end of the twig gets damaged, they may break it off to make the twig useable again. These operations on twigs need less effort and determination than knocking chips off a stone tool, but they may nevertheless be regarded as toolmaking. *Homo habilis* was similar in body size to chimpanzees and *Australopithecus*, but had a considerably larger brain. The volume of the braincase is about 0.6 liters, compared to 0.45 for *Australopithecus*.

*Homo erectus* ("erect man", not illustrated) is a little more advanced than *Homo habilis*. The Nariokotome boy, a remarkably complete skeleton from Kenya, shows that the shapes of the pelvis and rib cage, and the relative lengths of the arms and legs, were very much as in modern people. A few fossils of *Homo*, either *erectus* or a similar species, have been found among the *Paranthropus* at Swartkrans. There are also some antelope bones that have been damaged by fire. Experiments in which fresh antelope bones were heated in a campfire resulted in similarly altered bones. Here we have evidence of fire being used, but not necessarily of ability to make fire. The *Homo* seem more likely than the *Paranthropus* to have been responsible.

Lucy and the Nariokotome boy are famous fossils, but the Piltdown skull is notorious. [Plate 108] Its discovery in a gravel pit was announced in 1912. At that time, Java man (a *Homo erectus*) was the earliest known fossil hominid, and the Piltdown skull seemed to be of comparable age. Java man is merely a skull roof, but the Piltdown fossil has an

*Plate 109 (opposite) A cast of a Neanderthal skull. (Courtesy Division of Anthropology, American Museum of Natural History.)*

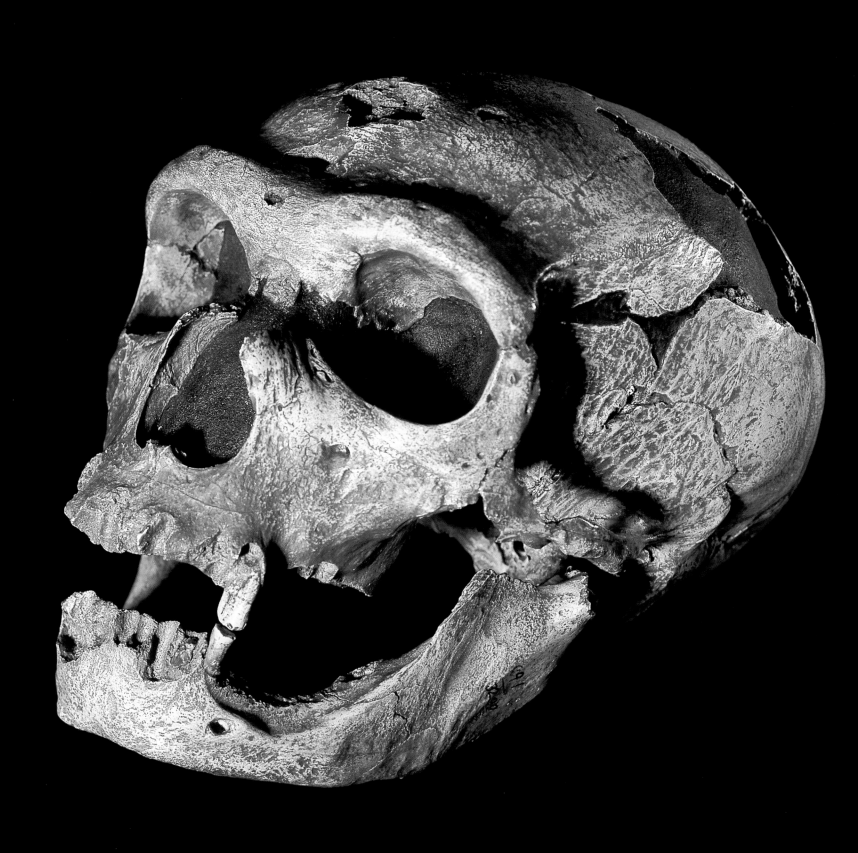

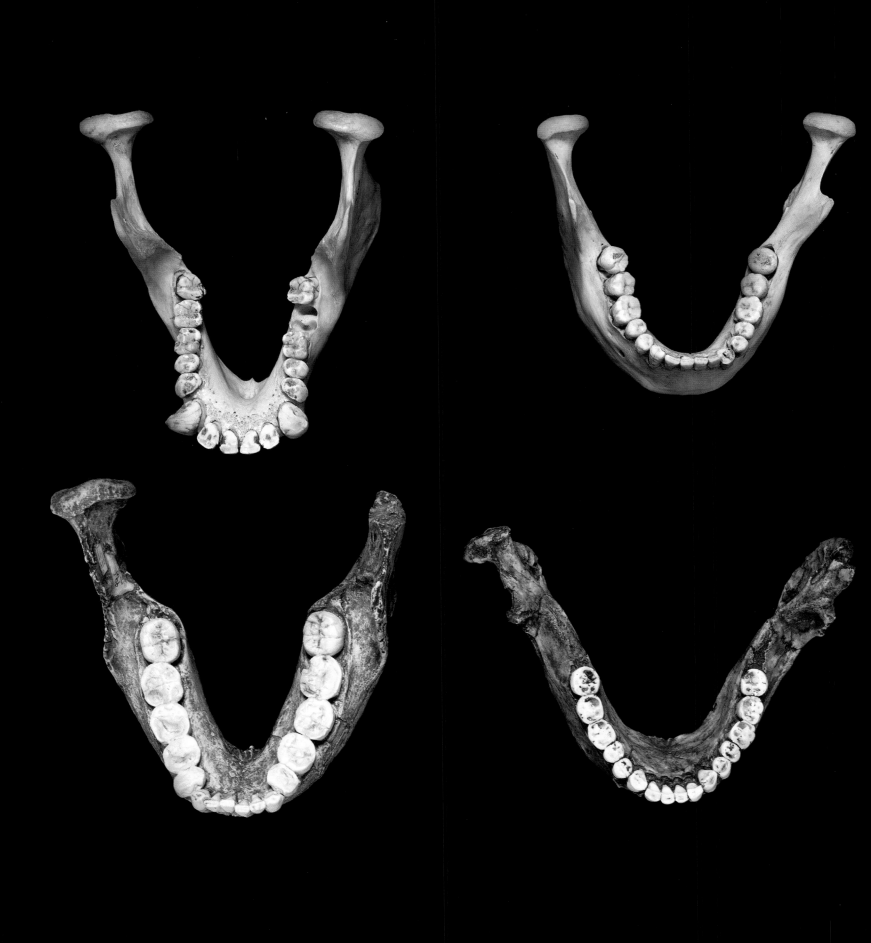

incomplete lower jaw and a few teeth as well as quite large pieces of the back of the braincase. These pieces showed that the brain was not much, if at all, smaller than modern human brains, but the jaw was distinctly ape-like. Together they seemed to show that our ancestors had evolved a big brain before the jaw had undergone much change.

The Piltdown skull soon aroused suspicions, but it was not until 1953 that it was shown conclusively to be a hoax. The skull fragments are mediaeval *Homo sapiens*, shown by several techniques to date from the fourteenth century. The jawbone came from an orangutan, and the molars are filed-down human ones. All the pieces had been stained brown with potassium dichromate to match the gravel in which they had been planted.

*Homo erectus* was the first hominid species to spread out from Africa. Specimens have been found in Indonesia (Java man is the most famous) and China (Peking man). From Asia, hominids spread to Europe, where *Homo neanderthalensis* seems to have evolved. [Plate 109] The first specimen was found in the Neander valley, in Germany: hence the name. The braincase is about the same size as in modern human skulls. The jaws do not project as far forward as in *Homo habilis* and *erectus*, but they project more than in *Homo sapiens*. The receding forehead and prominent brow ridges give the skull a distinctly primitive look. From the neck down the skeleton is heavily built but otherwise very like *Homo sapiens*.

Archaeological evidence shows that Neanderthals made much better tools than *Homo erectus*, with long cutting edges. Wear on some of the tools seems to be the result of cutting wood, perhaps sharpening wooden spears. The Neanderthals do not seem to have made stone spearheads. Bones of deer and other mammals found with Neanderthal remains seem to show that they ate meat. At some sites there are disproportionate numbers of leg bones, suggesting that Neanderthals cut off and carried home the meatiest parts of carcasses. It is uncertain whether they killed most of their meat, or scavenged on carcasses of animals that had died already. Hearths have been found, indicating that the Neanderthals had control of fire. They buried their dead, but there is no evidence that they made a ceremony of it. The few stone tools and other possessions that have been found in their graves may have got there by accident.

## NEANDERTHALS

It used to be thought that Neanderthals were our direct ancestors, but it now seems clear that *Homo sapiens* appeared independently in Africa, and spread out from there. Indeed, the Neanderthals seem to be only distantly related to us. Geneticists have been able to extract Neanderthal DNA from bones, and have compared it to modern human and chimpanzee DNA. They found an average of 27 differences in comparisons between

*Plate 110 (opposite) Lower jaws of (top left) chimpanzee; (lower left) Paranthropus boisei; (lower right) a Neanderthal and (top right) a modern human. (Courtesy Division of Anthropology, American Museum of Natural History.)*

Neanderthal and modern human DNA, and 55 between chimpanzees and modern humans. By this test, the relationship between them and us is only twice as close as between us and chimps.

*Homo sapiens* braincases are no bigger than Neanderthal ones, but their shape is different, giving us more upright foreheads and much less prominent brow ridges. Our jaws are shorter, relative to the rest of the skull, than the jaws of other hominids. This makes the face vertical from forehead to chin, with only the nose projecting forward.

A comparison of hominid jaws [Plate 110] shows that ours are U-shaped, whereas the jaws of apes and australopithecines are more V-shaped. Only *Homo sapiens* has a prominent chin; notice how it projects in front of the incisor teeth. Plate 89 (right) shows the passage through a modern woman's pelvic girdle, through which babies have to pass. It is bigger in humans and Neanderthals than in other hominids, and needs to be to let the head and its big brain through. Even so, being born is a very tight squeeze.

This chapter has shown that we have a reasonably coherent understanding of our evolution from apes, but information is still coming in. While I was writing this chapter the discovery was announced of *Sahelanthropus*, seven million years old, in Chad. Its significance will not be clear until it has been thoroughly studied. Is it a very early stage in the evolution of the hominids, or just an unusual ape?

We started this book by looking at bone as a living tissue that grows and repairs itself. The three chapters that followed reviewed the bones of the human skeleton, discussing their design and their functions. We went on to see how bones may be damaged by disease or by violence, and to consider the differences between the bones of different people. A recurring theme of the book has been that to understand the skeleton, we need to know about its evolutionary history, as well as its present-day functions. This last chapter has reinforced the message.

This exploration of the human skeleton has involved anatomy, physiology, mechanics, archaeology, anthropology, paleontology, orthopedics, dentistry and forensic science. And it has given us the opportunity to admire the beauty of the skeleton as displayed in Aaron Diskin's photographs. But science and aesthetics are not the only keys to appreciation of the skeleton. We finish with a reminder of an aspect of the skull that this book has ignored, as a potent symbol. [Plate 111]

*Plate 111 (opposite) An early twentieth century Tibetan kapala, a ritual bowl made from a human skull. This one has been fitted with turquoise eyes and decorated with gold leaf. (Courtesy Henry Galiano.)*

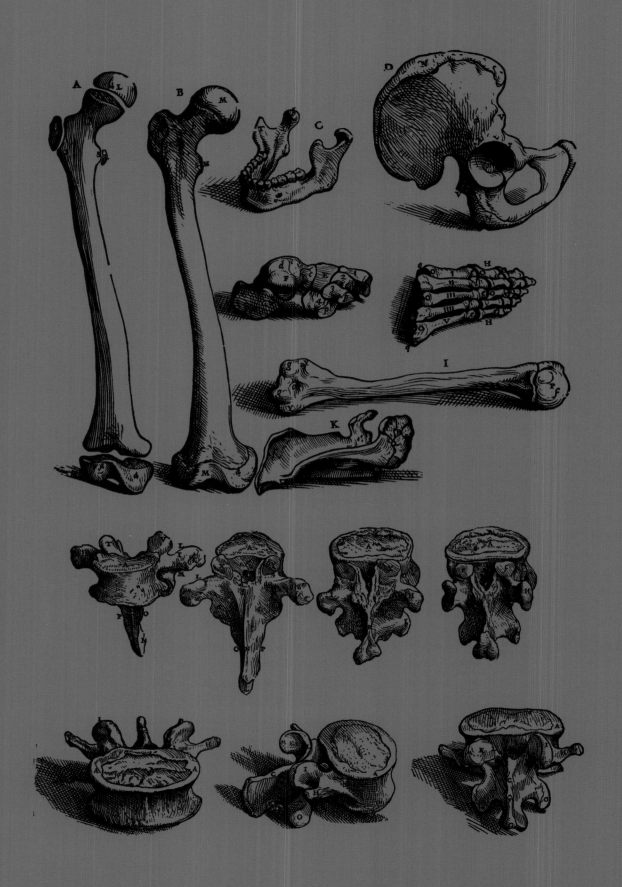

## LIST OF HUMAN BONES

7 cervical vertebrae

12 thoracic vertebrae

5 lumbar vertebrae

sacrum

coccyx

3 parts of sternum

24 ribs

mandible

5 parts of hyoid

occipital bone

sphenoid bone

2 temporal (squamosal) bones

2 parietal bones

frontal bone

ethmoid bone

2 inferior nasal conchae

2 maxillae

2 palatine bones

2 zygomatic (jugal) bones

2 lacrimal bones

2 nasal bones vomer

6 auditory ossicles

2 scapulas

2 clavicles

2 humerus

2 radius

2 ulnas

16 carpal bones (8 in each wrist)

10 metacarpals (5 in each hand)

28 phalanges (14 in each hand)

2 innominate (pelvic) bones

2 femurs

2 patellas

2 tibias

2 fibulas

14 tarsal bones (7 in each ankle)

10 metatarsal bones (5 in each foot)

28 phalanges (14 in each foot).

# PHOTOGRAPHER'S NOTE

When I was young, my Uncle Saul brought me a horse skull from Arizona. Long exposure to the sun had bleached and pitted the bone. It fascinated me, and I would spend hours tracing its lines with my fingers. It seemed like a magical object.

One winter when I was twenty, I bought two goat heads and a cow head from a butcher. They had been skinned, but were still sheathed in bloody muscle. Fresh, red, and glistening, they attracted me in an urgent way. I photographed the heads often and quickly in my snowy backyard.

When spring came, the flesh began to rot. I visited the putrefying heads once a week or so, my wonder becoming mingled with disgust: they smelled awful, and colonies of insects were feasting and living in and on the grisly craniums. I documented the decay, albeit sporadically.

One year later only bones remained. I brought the skulls into my living space, and I have been slowly photographing them for years. They hang from the wall above my head when I sleep. Without my loving and much beloved family, Vilunya Diskin, Leah Diskin, Tom Gaudette, and Nadia Gaudette-Diskin, my photography would not have been possible.

I would like to dedicate the work I did for this book to my father Martin Diskin.

*Aaron Diskin*
*Brooklyn, New York*

*Plate 112 (opposite) Skull and upper torso skeleton. (Courtesy of Maxilla & Mandible.)*

# CONSULTANT'S NOTE

I am passionate about natural history. It is this passion that prompted me to open Maxilla & Mandible in 1982, and encouraged me to participate in the making of *Human Bones*. In our bones, we can see humankind in all its diversity and imperfections. Every bone in our body reveals clues to the evolutionary history of humans on Earth. They are the only part of our bodies that can be readily compared to the anatomy of extinct humans and other animals. Man is forever pushing the borders of experience by intervening in the natural formation of bones. No other creature deforms its own body in pursuit of an aesthetic ideal, such as occurs with foot binding or cradle board molding of the skull. Human bones have always had an element of the taboo. I prefer to view them as representative of the greatness of the natural world.

Customers visiting Maxilla & Mandible sometimes comment about a medical skeleton on display, "Poor guy, why would you do this to a person?" Instead of being cremated or embalmed, I would rather end up as a useful skeletal specimen in a scientific collection or displayed for aesthetic appreciation. There is no dishonor in being preserved as a skeleton.

*Henry Galiano*
*Maxilla & Mandible*
*New York, New York*

*Plate 113 (opposite) Human skeleton.*
*(Courtesy of Maxilla & Mandible.)*

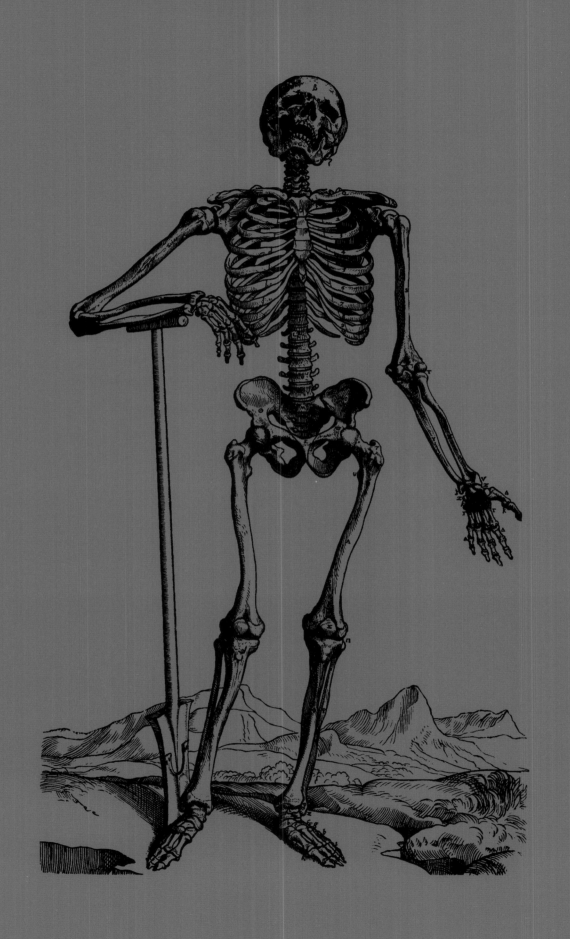

# ACKNOWLEDGMENTS

*Human Bones: A Scientific and Pictorial Investigation* is the result of the combined visions and efforts of scientists and artists, stretching from New York City (including Brooklyn) to Leeds, England.

First and foremost it is the result of the scholarship and wisdom of R. McNeill Alexander. Neill Alexander is arguably the repository of more knowledge and understanding about the vertebrate skeleton than anyone in the world today. The writing he has brought to *Human Bones* has a sure-footed grace that comes with 50 years of research and over 20 books and 250 journal articles on the form and function of vertebrates.

Aaron Diskin is a young, curious and eager photographer who has created for *Human Bones* a sequence of images that have the potential of changing how we perceive our own human bones. His dedication to the book was unwavering. Henry Galiano, proprietor of New York's legendary Maxilla and Mandible and connoisseur of natural history, provided many of the most interesting specimens pictured in *Human Bones* and shared crucial advice on how to most effectively present the human skeleton within the framework of this book. His partner, Deborah Galiano, was kind enough to lend for photography perhaps the most unusual specimen in the book: the skeleton of a bound foot seen on page 129. Elizabeth Murray helped organize the Maxilla and Mandible specimens for photography. Gary Sawyer, Senior Technician, Kenneth Mowbray, Curatorial Associate, and Ian Tattersall, Curator, in the Division of Anthropology, at the American Museum of History, generously made their entire collection of human and other vertebrate bones available for photography and gave the photographer, Aaron Diskin, much valuable advice and assistance during three weeks of intense photography at the Museum. Richard Monk, Curatorial Associate, and Nancy Simmons, Curator, in the Division of Mammalogy at the American Museum of Natural History provided important specimens from their collection.

Others who assisted in the photography include Tamara Reynolds, Peter Mauney, Laura Gail Taylor, Matt Pokoik, Alison Slagowitz, everyone who modeled: Gabriel Terizzi, Johnathan Chick, Christy Edwards, Lilah Freedland, Taylor Bergren-Chrisman, Seth Prouty, Dreiky Magana and William Haugh.

Matthew Schwartz, principle in Matthew Schwartz Design Studio, gave *Human Bones* its bold jacket and book design. Simone Nevraumont rendered the drawings that tastefully complement the photographs.

In the end, however, all this work would be for naught except for the commitment of Stephen Morrow, Executive Editor, of the Pi Press division of Pearson Technology Group, to publish *Human Bones*. He and his colleagues Logan Campbell, Dana Filippone and Jeff Galas believed in this book from the beginning and made it possible.

*Peter N. Nevraumont*
*Nevraumont Publishing Company*

# FURTHER READING

ALEXANDER, R. MCNEILL. *The Human Machine.* London: Natural History Museum Publications, and New York: Columbia University Press, 1992.

———*Bones: The Unity of Form and Function.* Boulder, Colorado: Westview Press, 2000.

BAHN, PAUL (ED.). *Written in Bones: How Human Remains Unlock the Secrets of the Dead.* Toronto: Firefly Books, 2002.

COLE, FRANCIS J. *A History of Comparative Anatomy: From Aristotle to the Eighteenth Century.* New York: Dover Publications, 1975.

CURREY, JOHN D. *Bones: Structure and Mechanics.* Princeton: Princeton University Press, 2002.

GEE, HENRY. *Before the Backbone: Views on the Origin of the Vertebrates.* London: Chapman and Hall, 1996.

GOLDFINGER, ELIOT. *Human Anatomy for Artists.* New York: Oxford University Press, 1991.

GOULD, STEPHEN JAY, and RICHARD C. LEWONTIN. "The Spandrels of San Marco and the Panglossian Paradigm: A Critique of the Adaptationist Program". *Proceedings of the Royal Society of London*, Series B, 205/1161 (1979): 581-598. Reprinted in ELLIOT SOBER (ED.) *Conceptual Issues in Evolutionary Biology.* Cambridge, Massachusetts: MIT Press, 1984.

JOHANSON, DONALD, and BLAKE EDGAR. *From Lucy to Language.* New York: Simon and Schuster, 1996.

MUYBRIDGE, EADWEARD. *The Human Figure in Motion.* New York: Dover Publications, 1955.

RENFREW, COLIN and PAUL BAHN. *Archaeology: Theories, Methods and Practice ed.3.* London: Thames and Hudson, 2000.

TATTERSALL, IAN. *The Fossil Trail: How We Know What We Think We Know about Human Evolution.* New York: Oxford University Press, 1995.

THOMPSON, D'ARCY. *On Growth and Form.* Cambridge: Cambridge University Press, 1917 (1st ed.), 1942 (2nd ed.).

VESALIUS, ANDREAS. *The Illustrations from the Works of Andreas Vesalius of Brussels.* New York: Dover Publications, 1973.

VON HAGEN, GUNTHER. *Visible Human Body: An Atlas of Sectional Anatomy.* Philadelphia: Lea and Febiger, 1990.

WHITE, TIM. *Human Osteology.* San Diego, California: Academic Press, 2000.